Active Tectonics
Earthquakes, Uplift, and Landscape

Edward A. Keller
University of California, Santa Barbara

Nicholas Pinter
University of California, Santa Barbara

Prentice Hall
Upper Saddle River, New Jersey 07458

Library of Congress Cataloging-in-Publication Data

Keller, Edward A.
      Active tectonics : earthquakes, uplift, and landscape / Edward
Keller, Nicholas Pinter.
         p.   cm.
      Includes bibliographical references and index.
      ISBN 0-02-363261-5 (pbk.)
      1. Plate tectonics. 2. Earthquakes. 3. Geomorphology.
I. Pinter, Nicholas. II. Title.
QE511.4.K45 1996
551.8—dc20   95-4854
                                                CIP

Acquisitions editor: Robert McConnin
Editor-in-chief: Paul Corey
Editor-in-chief of development: Ray Mullaney
Marketing manager: Leslie Cavaliere
Managing editor: Kathleen Schiaparelli
Director of production and manufacturing: David W. Riccardi
Manufacturing buyer: Trudy Pisciotti
Cover designer: Jayne Conte
Cover photo: Haruyoshi Yamaguchi/Sygma
Illustrator: David Crouch
Photo editor: Lorinda Morris-Nantz
Photo researcher: Terri Stratford
Production: Custom Editorial Productions, Inc.

© 1996 by Prentice-Hall, Inc.
Upper Saddle River, New Jersey 07458

Reprinted with corrections July, 1999

Printed in the United States of America

10 9 8 7 6 5 4 3 2

ISBN 0-02-304601-5

Prentice-Hall International (UK) Limited, *London*
Prentice-Hall of Australia Pty. Limited, *Sydney*
Prentice-Hall Canada, Inc., *Toronto*
Prentice-Hall Hispanoamericana, S. A., *Mexico*
Prentice-Hall of India Private Limited, *New Delhi*
Prentice-Hall of Japan, Inc., *Tokyo*
Prentice-Hall Asia Pte. Ltd., *Singapore*
Editora Prentice-Hall do Brasil, Ltda., *Rio de Janeiro*

To Professor Marie Morisawa, SUNY Binghamton, for her many contributions in geomorphology, her assistance and guidance of students, friends and colleagues, and her love of science and life.

Contents

7   Active Folding and Earthquakes   207

8   Paleoseismology and Earthquake Prediction   245

Preface

Active tectonics (also called **tectonic geomorphology***)* is the study of the dynamic Earth—the processes that occur, how those processes shape the landscape, and how those processes impact human society. In recent years, tectonic geomorphology increasingly has become one of the principal tools in a variety of applications including identification of active tectonic features, seismic hazard mapping, and understanding the development of the Earth's landscapes. Tectonic geomorphology has proven to be useful in these applications because landforms are created and preserved over time intervals ideal for recording the particular details of tectonic activity.

This book requires only basic knowledge of geologic principles. It is appropriate for upper division undergraduate students, graduate students, and others who work in the fields of geology, geomorphology, and earthquake studies. In universities, this book is appropriate for classes in tectonic geomorphology, natural hazards, and geomorphology. We also have designed the text to be highly applied, so that it will serve city planners, seismic engineers, and other non-geologists.

The field of active tectonics is evolving very rapidly. At the cutting edge of the field, space-based positioning and high-precision geodesy are bringing a whole new class of information to studies of the dynamic Earth. Advances in topics such as buried reverse faulting, river dynamics, climate change, and isostasy continue to refine and redefine our understanding of tectonic and geomorphic processes. We hope the readers of this book will find it to be an up-to-date source of information, as well as a solid foundation for understanding future advances in the field of active tectonics.

Acknowledgments

The authors would like to thank the reviewers who reviewed all or parts of the manuscript: Ronald L. Bruhn, University of Utah; Thomas W. Gardner, Pennsylvania State University; David R. Hickey, Graptolithics; William R. Lettis, Lettis & Associates, Inc.; Nancy Lindsley-Griffin, University of Nebraska-Lincoln;

George W. Moore, Oregon State University; Gomaa I. Omar, University of Pennsylvania; William A. Smith, Western Michigan University; and Steven N. Ward, University of California, Santa Cruz.

The authors are also pleased to acknowledge the assistance of editors Robert McConnin, Tim Flem, Ed Thomas, and Diane Sparks. Assistance from Ellie Dzuro (word processing), Amy Wiltse (administrative assistance and editing), Dave Crouch (computer illustration), and Christopher Rogers (production) is also greatly appreciated.

1
Introduction to Active Tectonics: Emphasizing Earthquakes

ACTIVE TECTONICS

In the geological sciences, the term "tectonics" refers to the processes, structures and landforms associated with deformation of the Earth's crust. In a broader sense, it refers to the evolution of these structures and landforms over time. On a global scale, we are concerned with the origin of continents and ocean basins, which are the largest landforms produced by tectonics on Earth. At the regional scale, we are interested in structures that produce mountain chains. At the local scale, we often study features such as small folds, which may form elongated low hills, or faulting, which may result in ground rupture and the production of relatively small (a few tens to thousands of meters long by less than 1 m to about 8 m high), steep slopes known as fault scarps.

The time scales of tectonics depend on the spatial scale at which the processes act. For example, it takes billions of years for continents to develop; hundreds of millions of years for large ocean basins; several million years for small mountain ranges; several hundred thousand years for small folds to produce hills; and fault scarps may be produced almost instantaneously during earthquakes.

Rates of tectonic process are also extremely variable.

- Fault ruptures during earthquakes may propagate as fast as several kilometers per second. For example, during the 1983 Borah Peak earthquake in Idaho, eyewitnesses reported that a fault scarp approximately 1 m in height formed in less than 1 sec [1].
- When tectonic processes are averaged over a period of years, rates generally range from fractions of millimeters to several millimeters per year for fault displacement to several centimeters per year for processes that move continents and form ocean basins at locations where new oceanic crust is being produced, where old oceanic crust is being consumed, or where plates are colliding.

The term "active tectonics" refers to those tectonic processes that produce deformation of the Earth's crust on a time scale of significance to human society [1]. As such, we are most interested in processes likely to cause disruption of society within a period of several decades to several hundred years—the time period for which we plan the lifetimes of buildings and important facilities such as dams and power plants. However, in order to study and predict tectonic events over this time period, we must study these processes over a much longer time scale—at least several thousand years to several tens of thousands of years—because earthquakes on particular faults may have long return periods (time between events). Depending on when the most recent event occurred, faults may produce earthquakes in the next several decades or the next several thousand years.

Another viewpoint is that the time frame necessary to study active tectonics is more like several millions of years [2]. The argument is that present or contemporary tectonic activity and associated deformation may be partially or predominantly controlled by the tectonic framework (geometry and mode of operation of structures such as faults and folds) of the previous 500,000 yr. According to this viewpoint, understanding the tectonic processes over several millions of years is necessary to fully understand active tectonics and mitigate associated geologic hazards such as earthquakes [2].

Although active tectonics includes the slow disruption (warping or tilting) of the Earth's crust that may cause damage to human structures, we are most concerned with active tectonic processes capable of producing catastrophes. A **catastrophe** is defined as any situation in which the damage to people, property, or society in general is sufficiently severe that recovery or rehabilitation, or both, are a long, involved process [3]. One active tectonic process likely to produce a catastrophe is a great earthquake. However, moderate-sized earthquakes also can produce catastrophes, particularly if they occur in densely populated areas where buildings are constructed of materials that cannot withstand shaking (homes constructed of unreinforced cement blocks, bricks, or stones are particularly hazardous) or buildings constructed on thick layers of unconsolidated sediment (particularly those sediments with a high water content). Catastrophic earthquakes in history include:

- A sixteenth-century event in China that reportedly claimed approximately 850,000 lives
- A 1923 earthquake near Tokyo that claimed 143,000 lives
- A 1976 earthquake in China that claimed several hundred thousand lives
- A 1985 earthquake originating in rocks below the Pacific Ocean off Mexico that sent seismic waves to Mexico City (several hundred kilometers from the source) and caused approximately 10,000 deaths
- A 1989 earthquake on the San Andreas fault system south of San Francisco, California, that killed 62 people and caused $5 billion in property damage
- The 1994 Northridge earthquake in the Los Angeles, California, urban area that killed 61 people and caused more than $20 billion in damage
- The 1995 Kobe, Japan, earthquake that killed about 5000 people and caused more than $30 billion in damages

It has been estimated that a great earthquake in a densely populated part of Southern California could do $100 billion in damage and kill several thousand people [4]. Thus, the 1994 event, as terrible as it was, was not "the big one." Because earthquakes have the proven potential for catastrophic damage, much of our research in active tectonics is to better understand earthquake processes, the damage likely to occur, and the ways to minimize loss of life and property damage.

Research in active tectonics at a variety of scales, from regional to local, can be useful to society. Figure 1.1 is a generalized flowchart from data input (at the regional or local levels) to output and possible social impacts of active-tectonic information such as improving regional planning and site-specific land use, establishing building codes, and planning for earthquake-hazard reduction. At the data input stage, we make measurements and observations from topographic maps, aerial photographs, and field work that help define relative tectonic activity and areas where more detailed work is needed to better understand the earthquake hazard. This reconnaissance-level work delineates areas where detailed evaluation may provide rates of active-tectonic processes, recurrence intervals of earthquakes, and rates of crustal deformation. This information is necessary for society to develop regional planning strategies and site-specific strategies for earthquake-hazard

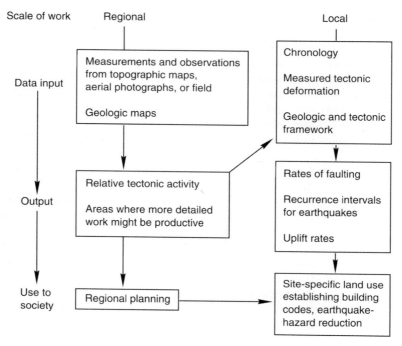

FIGURE 1.1
Active tectonics data input, output, and use to society.

reduction. It is important to keep in mind that the role of the earth scientist is to evaluate and predict types and rates of active-tectonic processes, and it is the role of engineers to design our homes, buildings, and other structures to better withstand these processes. However, it is the role of the policy makers to evaluate risk, develop management strategies, and implement policies that minimize loss of lives and property (that is, reduce the hazard level to an acceptable risk) [1]. With this general introduction, the remainder of this chapter will discuss selected aspects of earthquakes and related phenomena at global to local scales.

GLOBAL TECTONICS

Tectonic processes are driven by forces deep within the Earth that deform the crust, producing external forms such as ocean basins, continents, and mountains. Collectively these processes are known as the **tectonic cycle.** Earth scientists, through detailed study of the ocean basins and continents, have established that the **lithosphere**—the outer layer of the Earth which contains the continents and ocean basins—is relatively strong and rigid compared to deeper material, and ranges in thickness from several tens of kilometers beneath parts of ocean basins to greater than 100 km beneath parts of continents. The lithosphere is not a continuous uniform shell, but is broken into several large pieces called lithospheric plates that move relative to one another (Figure 1.2) [5]. Plates can include both continents and portions of ocean basins, or ocean basins alone. Some of the largest plates are the Pacific, North American, South American, Eurasian, African, and Australian. There are also numerous smaller plates that are significant at the regional scale, including the Juan de Fuca Plate off the Pacific northwest coast of the United States and the Cocos Plate off Central America (Figure 1.2).

The lithospheric plates move over the **asthenosphere,** which is thought to be a more-or-less continuous, hot, and plastically flowing layer of relatively weak rock below the lithosphere. This motion causes the continents to change their relative positions on the surface of the Earth (Figure 1.3) [6]. The idea that continents move is not new, but it is only in the last 25 years that this hypothesis has been accepted and studied intensively enough to gain the status of a unified theory of the Earth. Alfred Wegener, a German scientist, first suggested in the early twentieth century that the continents were moving or drifting. His evidence was based in part on the good fit of the continents, such as between South America and Africa, but most importantly on the similarities in rock types, geologic structures, and paleontological (fossil) evidence now found on opposite sides of the Atlantic Ocean. However, it was not until the late 1960s when the process of sea-floor spreading (shown on Figure 1.3) was discovered that a plausible *mechanism* for continental drift was provided. The most recent global episode of continental drift and sea-floor spreading started about 200 Ma, when Wegener's hypothesized supercontinent called Pangea broke up. Figure 1.4 shows Pangea as it was 200 Ma as well as the present positions of the continents and ocean basins. Sea-floor spreading in the past 200 m.y. separated Africa and Eurasia from North and South America, South America from Africa and Antarctica, and Australia and India from Antarctica. The Tethys Sea closed, leaving the small remnant of it today known as the Mediterranean Sea. About 50 Ma, India crashed into Eurasia, producing the Himalayan Mountains and Tibetan Plateau. That collision is still happening today.

The boundaries between lithospheric plates are areas of geological activity where most earthquakes and volcanic activity occur. The three types of boundaries between the plates are divergent, convergent, and transform [7]. **Divergent boundaries** occur at spreading ridges where new lithosphere is produced and plates move away from each other (Figure 1.3). **Convergent boundaries** occur where one plate dives ("subducts") beneath the leading edge of another plate, and thus are also known as subduction zones. However, if both leading edges are composed of relatively low-density continental material (average composition of granite), it is more difficult for subduction to start, and a special type of convergent plate boundary, called a **continental-collision boundary,** may develop. This type of boundary condition produces some of the highest linear mountain systems on Earth, such as the Alps and the Himalayas. **Transform boundaries** occur where one plate slides past another, displacing spreading ridges. This type of boundary is most common in oceanic crust, but it also occurs on land, as for example along the

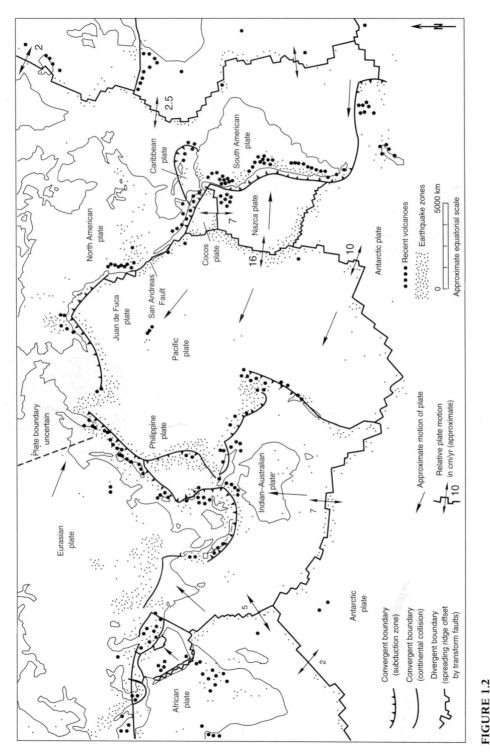

FIGURE 1.2

Lithospheric plates forming the Earth's outer layer. Three types of plate junctions are shown: spreading ridges, forming divergent boundaries; subduction zones, forming convergent plate boundaries; and, more rarely, transform plate boundaries, such as the San Andreas fault in California, where one plate is sliding by another.

(After Bolt, 1988 [9]. Rates after Minster and Jordan, 1988. *Journal of Geophysical Research,* 83: 5331–5354)

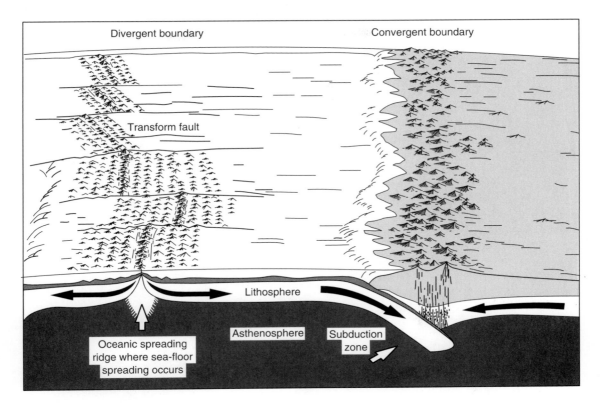

FIGURE 1.3
Diagram of the model of sea-floor spreading which drives the movement of lithospheric plates. New lithosphere is being produced at the spreading ridge (divergent plate boundary). The lithosphere then moves laterally and eventually returns down to the interior of the earth at a convergent plate boundary (subduction zone). This process produces ocean basins and provides a mechanism that moves continents.
(After Lutgens and Tarbuck, 1992. *Essentials of Geology.* Macmillan: New York)

San Andreas fault in California (Figure 1.2). At some locations, three plates meet, and these areas are known as **triple junctions.** One example of a triple junction is where the Juan de Fuca, North American, and Pacific plates meet (Figure 6.1 is a detailed map of this area). Another example of a triple junction is located west of South America north of the equator (Figure 1.2), where the three spreading ridges between the Pacific, Cocos, and Nazca plates meet.

Rates of plate motion are shown on Figure 1.2. In general, the rates are about as fast as fingernails grow, but vary from about 2 cm/yr to 15 cm/yr. The San Andreas fault moves at a rate of about 3.2 cm/yr on average. The displacement is termed right-lateral strike-slip because the displacement is predominantly horizontal, and features such as rock units, streams, or chain link fences are displaced to the

A

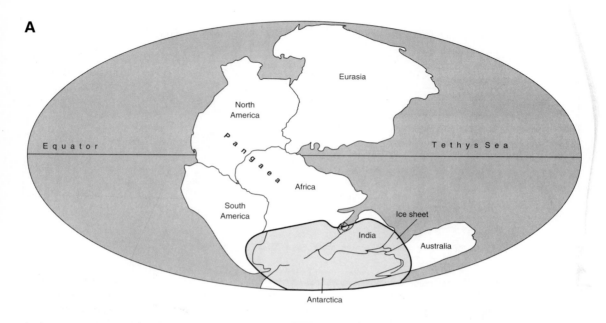

B

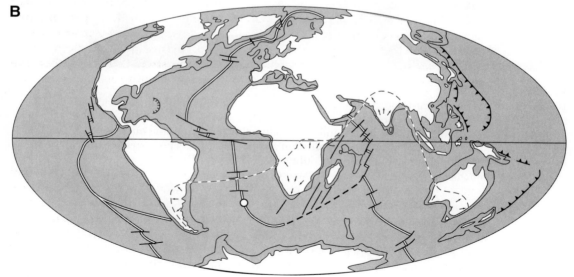

FIGURE 1.4
(A) Pangaea at 200 Ma and (B) present position of continents. Notice that the polar glaciation in (A) left evidence in five continents (B). The arrows show direction of ice movement
and only make sense if the continents are moved back to their position 200 Ma.
(After Skinner and Porter, 1989. *The Dynamic Earth*. John Wiley & Sons: New York)

right where they cross the fault. During the past 4 m.y., there has been about 200 km of displacement on the San Andreas Fault. Los Angeles, which is on the Pacific Plate, is slowly moving toward San Francisco (which is on the American Plate), and in 10–20 m.y. the cities will be side by side. Movement at plate boundaries is not necessarily a steady process from year to year, but rather may occur in episodic jumps during great earthquakes. Fortunately, such events only occur in any one location every few hundred years.

The good correlation between plate boundaries and earthquakes is dramatically shown on Figure 1.5, which is a map of global seismicity from 1963 to 1988. The locations of earthquakes clearly are related to the major plate boundaries and, in fact, can be used to map the boundaries. It is important to recognize that several large and damaging earthquakes also have occurred far from plate boundaries. However, these events are the exception and not the rule.

EARTHQUAKES AND RELATED PHENOMENA

Our discussion of global tectonics established that the Earth is a dynamic, evolving system. Earthquakes are a natural consequence of the dynamic processes forming the ocean basins, continents, and mountain ranges of the world. As new lithosphere is produced at oceanic ridge systems, older lithosphere is consumed at subduction zones, or as plates slide past one another. **Stress** (force per unit area on a specified plane, in a material such as rock) is produced, and **strain** (deformation, such as change in length or volume, or rupture resulting from stress) builds up in the rocks. When the stress exceeds the strength of the rocks, the rocks fail (rupture) and energy is released in the form of an earthquake. As a result, faults are considered **seismic (earthquake) sources.** Identification of seismic sources in an area is the first step in evaluating the earthquake risk.

The process of **faulting** can be compared to sliding two rough boards past one another. Friction along the boundary between the boards may temporarily slow their motion, but rough edges break off and motion occurs along the boundary. This process is analogous to what happens at plate boundaries where one plate slides past or overrides another. The rocks undergo strain and, if the stress continues, the rocks eventually break, forming a **fault.** A fault is defined as a fracture or fracture system along which rocks have been displaced; that is, rocks on one side of the fault have moved relative to rocks on the other side. Figure 1.6 shows the major types of faults based upon sense of relative displacement. Figure 1.7 shows selected aspects of the three types of faults—**normal, thrust** (a very low angle **reverse** fault), and **strike-slip**—and how they may look at the surface. Chapter 2 provides a detailed discussion of how faulting shapes the landscape. Most of the faults shown on Figure 1.6 displace the surface. However, some faults are buried; that is, fault rupture during earthquakes does not propagate to the surface even during large earthquakes (as for example, the 1994 Northridge earthquake). Buried faults are commonly associated with folding of rock. This important class of faults is discussed further in Chapter 7 on active folding.

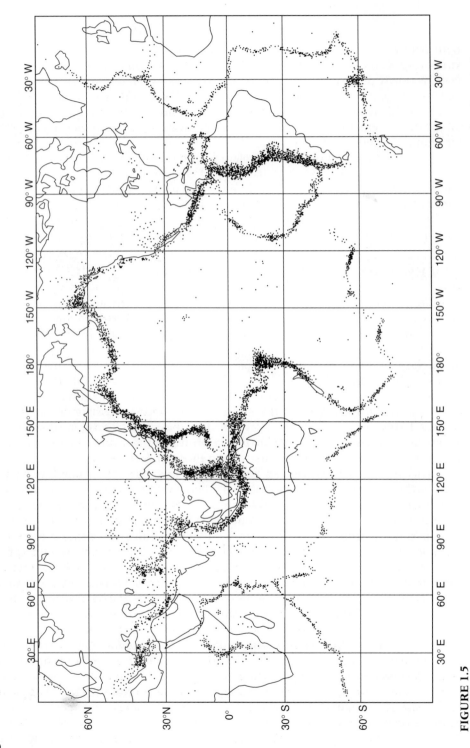

FIGURE 1.5

Map of global seismicity (1963–1988, Richter magnitude $M \geq 5$), delineating belts of earthquake activity that define plate boundaries. Compare with Figure 1.2.

(Courtesy of National Earthquake Information Center)

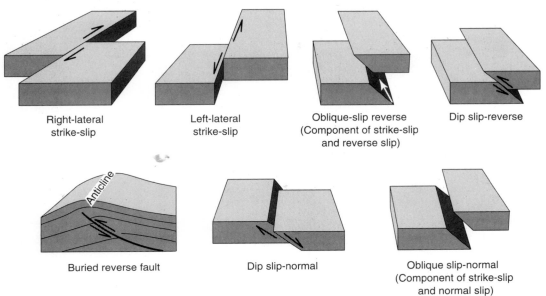

FIGURE 1.6
Types of fault movement based on the sense of fault motion.
(Modified from Wesson et al., 1975. *U.S. Geological Survey Professional Paper* 941A)

Faults almost never occur as a single trace or rupture. Rather, they form fault zones. A **fault zone** is a group of related fault traces that are subparallel in map view and often partially overlap in en echelon or braided patterns. Fault zones vary in width, ranging from a meter or so to several kilometers wide.

Most long faults such as the San Andreas fault are **segmented,** each segment having an individual style and history, including earthquake history. Rupture during an earthquake is thought to stop at the boundaries between two segments; however, great earthquakes may involve several segments of a fault zone. When the earthquake history of a fault zone is unknown, individual fault segments may be characterized based on changes in fault-zone morphology or geometry. It is preferable, from an earthquake-hazard evaluation point of view, to segment faults based on seismic activity and paleoseismic evaluation (see Chapter 8). Considerable research is being conducted to better understand the geology and processes that govern fault segmentation and the earthquakes generated on individual segments. Fault segmentation is discussed in later chapters.

Because most earthquakes are concentrated near plate boundaries, most large U.S. earthquakes are in the West, particularly near the North American and Pacific plate boundaries. However, large, damaging earthquakes also have occurred far from plate boundaries; these are termed **intraplate earthquakes.** For example, during the winter of 1811–1812, a series of particularly strong earthquakes struck the central Mississippi Valley, nearly destroying the town of New Madrid, Missouri, and killing an unknown number of people. These earthquakes rang

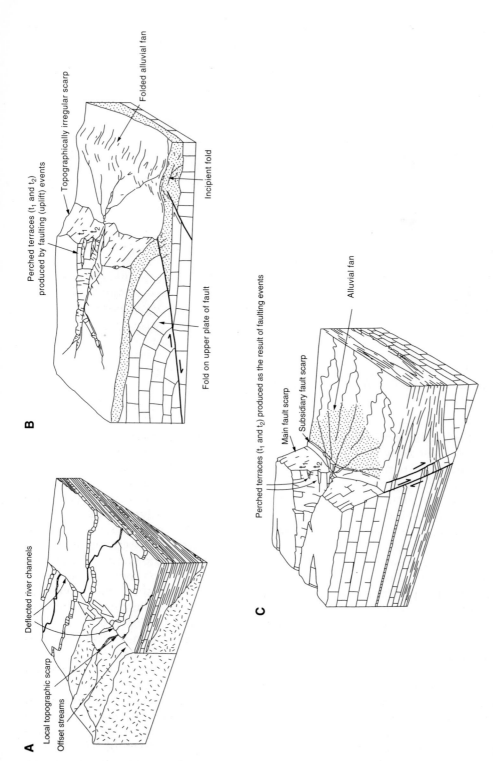

FIGURE 1.7

Idealized diagrams showing types of topographic expression possible with different types of faults. (A) A strike-slip fault showing a local topographic scarp, deflected river channels, and offset streams. (B) Thrust fault (low-angle reverse fault) with a fold on the upper fault plate. Topography shows an irregular scarp, perched terraces (t_1 and t_2) produced by faulting (uplift) events, and a folded alluvial fan. (C) Normal fault, with topography showing main and subsidiary fault scarps, perched terraces (t_1 and t_2) produced by faulting (uplift) events, and an alluvial fan. (After Ramsen and Huber, 1987. *Modern Structural Geology*, Vol. 2. Academic Press: New York)

church bells in Boston, over 1600 km away, and produced intensive surface deformation over a wide area from Memphis, Tennessee, north to the confluence of the Mississippi and Ohio Rivers. During the earthquakes, forests were flattened, fractures opened so wide that people had to cut down trees to cross them, and the land sank several meters in some areas, causing flooding. It was reported that the Mississippi River actually reversed its flow during the shaking [8]. The earthquakes occurred along a seismically-active structure known as the New Madrid Fault Zone, which underlies the geologic structure known as the Mississippi Embayment. The Mississippi Embayment is a downwarped rift in the crust where the lithosphere is relatively weak; it has broken repeatedly because compressional stress is transmitted from the distant boundaries of the plate. The recurrence interval for large earthquakes along the embayment is estimated to be 600–700 yr [8]. The possibility of future damage demands that the earthquake hazard be considered in design and construction of facilities such as power plants and dams, even in the "stable" interior of the North American Plate. A second example of a large intraplate earthquake is the 1886 event that nearly destroyed Charleston, South Carolina, taking 60 lives and causing $23 million in property damage.

MAGNITUDE AND INTENSITY OF EARTHQUAKES

The point within the Earth where earthquake rupture starts is called the **focus** [9]. The **epicenter** is the point on the surface of the Earth directly above the focus (Figure 1.8). News media usually report the location of the epicenter, but scientific reporting includes the location of the epicenter and the depth to the focus.

The depth of an earthquake's focus may vary from a few kilometers to almost 700 km. The deepest earthquakes occur along subduction zones, where slabs of brittle oceanic lithosphere sink to great depths. However, most earthquakes are relatively shallow, with foci less than about 60 km. In Southern California, most earthquakes have foci of about 10 km to 15 km depth, although deeper earthquakes do occur. The magnitude 7.4 event in 1992 on strike-slip faults in the Mojave Desert near Landers, California, had a focus of less than 10 km. The Landers event caused extensive ground rupture for about 85 km, with very local vertical displacement exceeding 2 m and extensive lateral displacements as much as about 5 m [10] (Figure 1.9). If the Landers earthquake had occurred in the Los Angeles Basin, it would have caused extensive damage and loss of life.

The **Richter magnitude** *(M)* of an earthquake (also known as the **local magnitude,** or M_L) is a measure of the amount of energy released and is useful in comparing earthquakes [9]. The Richter magnitude was first determined by the largest amplitude (in thousandths of millimeters) of seismic waves recorded on a standard **seismograph** (an instrument for recording earthquake waves) at a distance of 100 km from the earthquake epicenter [9]. The amplitude of the shaking is converted to a Richter magnitude using a logarithmic scale; for example, a $M = 7$ earthquake produces a displacement on the seismograph 10 times larger than does a $M = 6$. The energy released in an earthquake is proportional

FIGURE 1.8
Block diagram of a fault
plane and rupture area associated with an earthquake.
Also shown are the focus,
epicenter, and spreading rupture. When the rupture
reaches the surface, a fault
scarp is produced.

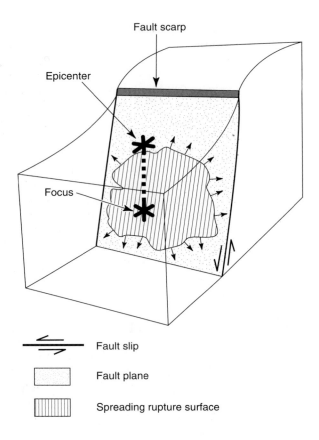

to the magnitude, but a one-unit change in Richter magnitude increases the
energy released by about 30 times. For example, a $M = 5$ earthquake releases
about 30 times more energy than a $M = 4$ event. Therefore, about 27,000 (30 ×
30 × 30) shocks of $M = 5$ are required to release as much energy as a single
earthquake of $M = 8$. Considering the entire Earth, there are about one million
earthquakes per year that can be felt by people. However, only a small percen-
tage of these can be felt very far from their source.

In a general sense, the Richter magnitude also may be related to the damage
expected from an earthquake. A $M = 8$ or above is considered to be a great earthquake, capable of causing catastrophic damage; an earthquake with Richter magnitude greater than 7 is a major earthquake, capable of causing widespread damage;
an earthquake of $M = 6$ can cause considerable damage, depending upon factors
such as location and surface materials present.

The Richter magnitude of an earthquake can also be estimated using graphical solutions as illustrated on Figure 1.10A. The maximum amplitude and difference in arrival time of P (primary) and S (secondary) waves from a distant
earthquake are measured from a seismograph. The example on Figure 1.10

FIGURE 1.9
Photograph showing ground
rupture from the Landers
earthquake, 1992.
Photograph by E.A. Keller.

shows the seismic record of an earthquake with an amplitude of 85 mm and a difference in arrival time (*S* minus *P*) of 34 sec. The line connecting the amplitude and difference in arrival time indicates the magnitude is 6, and the distance from the epicenter is about 300 km. Records from a minimum of three seismographs in a region are necessary to precisely locate the epicenter (Figure 1.10B).

The Richter magnitude scale was initially intended as a local magnitude (M_L) and the seismic wave used was the largest regardless of type (*P, S,* or surface). If the magnitude is determined using the largest amplitude of the *P* waves, then the resulting magnitude is termed M_b (the *b* stands for body wave, and *P* waves are a type of body wave; see Figure 1.11), and if the largest amplitude of a surface wave is used, then the magnitude is called the M_s.

In recent years there has been a move to change from the Richter magnitude to the **moment magnitude** scale. The moment magnitude is considered by seismologists to be a natural progression to a more quantitative and physically-based scale. The moment magnitude scale is based on the **seismic moment,** which is defined as the product of: (1) the average amount of slip on the fault that produced the earthquake, (2) the area that actually ruptured, and (3) the

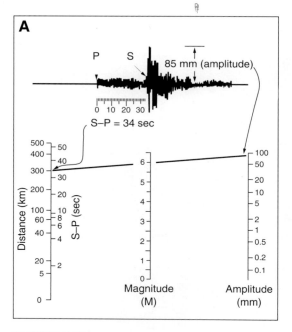

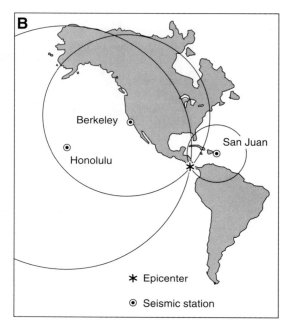

FIGURE 1.10

(A) Idealized diagram showing one procedure for determining the Richter magnitude (*M*) of an earthquake. For our example, the maximum amplitude (85 mm) is measured from the seismic record; the difference in arrival time between the *S* and *P* waves (34 sec) is also taken from the seismic record; the approximate magnitude of the earthquake as well as distance from the recording station is obtained by placing a straight line between the amplitude in millimeters and difference in arrival time in seconds, as shown on the diagram. Here, *M* = 6 and the distance is approximately 300 km. (B) Generalized concept of how the epicenter of an earthquake is located. Distance to event from at least three seismic stations is determined (Figure 1.10A) and plotted. The intersection of the arc distances (circle radii) defines the epicenter. For the diagram, the epicenter in Central America is located from data supplied by three seismic stations. Accurate location of the epicenter is not always as simple as the hypothetical example.
[(A) After Bolt, 1988 (9)]

shear modulus (resistance of a material to distortion by shear stress) of the rocks that failed [11]. In practice, seismic moment may be estimated for an earthquake by examining the records from seismographs, determining the amount and length of rupture, and estimating the shear modulus (strength) of rocks at the fault. The moment magnitude *(M_w)* is then determined from the mathematical relationship

$$M_w = 2/3 \log M_o - 10.7 \qquad (1.1)$$

where M_w is the moment magnitude, M_o is the seismic moment, and 10.7 is a constant [11]. The moment magnitude scale has a more sound physical basis and is

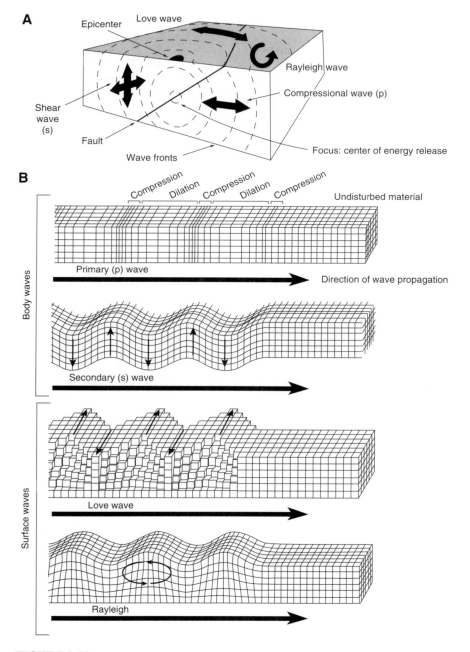

FIGURE 1.11

(A) Diagram of directions of vibration of body (*P* and *S*) and surface waves (Love and Rayleigh) generated by an earthquake associated with the illustrated fault. Also shown are the focus (center of energy release) and epicenter of the earthquake event. (B) Propagation of body and surface waves.

[(A) From Hays, 1981 (12); (B) After Bolt, 1988 (9)]

applicable over a wider range of ground motions than is Richter magnitude. Therefore, its use has been encouraged in reporting earthquake statistics.

A qualitative way of comparing earthquakes is to use the **Modified Mercalli Scale,** which is based on observations concerning the severity of shaking at different locations during an earthquake (Table 1.1). The scale has 12 divisions of intensity and, whereas a particular earthquake has only one Richter magnitude, the Modified Mercalli intensity at a particular location depends on its proximity to the epicenter and local geologic conditions. Questionnaires sent out to residents in the epicentral region after an earthquake are used to draw maps that show the spatial (areal) variability of intensity (see Table 1.2). Thus, the magnitude of an earthquake provides information concerning the amount of energy released, whereas intensity reflects how people perceived and how structures responded to the shaking. An example of the relationships between earthquake magnitude and intensity of shaking (Modified Mercalli Scale) for the 1971 ($M = 6.6$) San Fernando, California, earthquake is shown on Table 1.2. Also listed are values of **average peak horizontal acceleration** of the ground expected from earthquakes of $M = 3$ to $M = 9$. Estimated peak acceleration (both horizontal and vertical) is necessary information for designing buildings and other structures to withstand seismic shaking. Although the data on Table 1.2 suggest a good relationship between earthquake magnitude and average peak horizontal acceleration, the correlation is not nearly so good for the accelerations that occur close to the epicenter area. For example, during the $M = 6.6$ San Fernando earthquake of 1971, the maximum peak horizontal acceleration recorded at one site was 1.15g (g is the acceleration of gravity, equal to 9.8 m/sec^2). It is the horizontal acceleration that is most likely to cause damage to buildings. The 1994 Northridge earthquake locally produced even higher values of horizontal acceleration in the Los Angeles urban area. Unreinforced buildings constructed of adobe, common in Mexico, South America, and the Middle East, can collapse under a horizontal acceleration as small as 0.1g [9].

The vulnerability of unreinforced, prefabricated five-story apartments to seismic shaking was tragically illustrated on May 28, 1995, when a $M = 7.5$ earthquake struck the town of Neftegorsk on the island of Sakhalin off the northeast coast of Russia. Seventeen five-story structures, constructed without consideration of potential earthquake hazard in the late 1960s as part of a development to support oil production, collapsed into rubble. Approximately 2000 of the town's 3000 people were killed. The town appears to be a total loss and may not be rebuilt.

SEISMIC WAVES

Faulting breaks rock, and that movement along faults produces seismic waves that cause the ground to vibrate. Some of the seismic waves produced travel within the Earth and are known as **body waves,** whereas others travel along the surface (Figure 1.11). The two types of body waves generated by earthquakes are the primary, or *P* **waves,** and the secondary, or *S* **waves.** The *P* waves are the fastest of the two and, like sound waves, may travel through both solid and liquid materials. These waves push and pull in the direction of wave propagation with an alternating compression and dilation motion. The rate of propagation for *P* waves through rocks

TABLE 1.1
Modified Mercalli intensity scale (abridged).

Intensity	Effects
I	Not felt except by a very few under especially favorable circumstances.
II	Felt only by a few persons at rest, especially on upper floors of buildings. Delicately suspended objects may swing.
III	Felt quite noticeably indoors, especially on upper floors of buildings, but many people do not recognize it as an earthquake. Standing motor cars may rock slightly. Vibration like passing of truck. Duration estimated.
IV	During the day felt indoors by many, outdoors by a few. At night some awakened. Dishes, windows, doors disturbed; walls make cracking sound. Sensation like heavy truck striking building; standing motor cars rocked noticeably.
V	Felt by nearly everyone; many awakened. Some dishes, windows, etc., broken; a few instances of cracked plaster; unstable objects overturned. Disturbance of trees, poles and other tall objects sometimes noticed. Pendulum clocks may stop.
VI	Felt by all; many frightened and run outdoors. Some heavy furniture moved; a few instances of fallen plaster or damaged chimneys. Damage slight.
VII	Everybody runs outdoors. Damage negligible in buildings of good design and construction; slight to moderate in well-built ordinary structures; considerable in poorly built or badly designed structures; some chimneys broken. Noticed by persons driving motor cars.
VIII	Damage slight in specially designed structures; considerable in ordinary substantial buildings with partial collapse; great in poorly built structures. Panel walls thrown out of frame structures. Fall of chimneys, factory stacks, columns, monuments, walls. Heavy furniture overturned. Sand and mud ejected in small amounts. Changes in well water. Disturbs persons driving motor cars.
IX	Damage considerable in specially designed structures; well-designed frame structures thrown out of plumb; great in substantial buildings, with partial collapse. Buildings shifted off foundations. Ground cracked conspicuously. Underground pipes broken.
X	Some well-built wooden structures destroyed; most masonry and frame structures with foundations destroyed; ground badly cracked. Rails bent. Landslides considerable from river banks and steep slopes. Shifted sand and mud. Water splashed (slopped) over banks.
XI	Few, if any, masonry structures remain standing. Bridges destroyed. Broad fissures in ground. Underground pipelines completely out of service. Earth slumps and land slips in soft ground. Rails bent greatly.
XII	Damage total. Waves seen on ground surfaces. Lines of sight and level distorted. Objects thrown upward into the air.

[From Wood and Neuman, 1931. U.S. Geological Survey, 1974. *Earthquake Information Bulletin,* 6(5): 28]

TABLE 1.2

Approximate relationships between the magnitude and intensity of an earthquake, with the San Fernando Valley earthquake as an example.

Magnitude	Area Felt Over (square kilometers)	Distance felt (kilometers)	Intensity (maximum expected Modified Mercalli)	Ground Motion: (Average peak horizontal acceleration g = gravity = 9.8 meters per second per second)
3.0–3.9	1,950	25	II–III	Less than 0.15 g
4.0–4.9	7,800	50	IV–V	0.15–0.04g
5.0–5.9	39,000	110	VI–VII	0.06–0.015g
6.0–6.9	130,000	200	VII–VIII	0.15–0.30g
7.0–7.9	520,000	400	IX–X	0.50–0.60g
8.0–8.9	2,080,000	720	XI–XII	Greater than 0.60g

1971 San Fernando Valley Earthquake (M = 6.6)

Modified Mercalli intensity map

[After U.S. Geological Survey, 1974. *Earthquake Information Bulletin,* 6(5): 28]

such as granite is approximately 5.5 km/sec. The rate is much reduced through liquids; for example, *P* waves travel about 1.5 km/sec through water. Interestingly, *P* waves with frequency greater than about 15 Hz (cycles per second) are detectable to the human ear when propagated into the atmosphere, explaining why people sometimes hear earthquakes before feeling the shaking from the slower surface waves. *S* waves travel only through solid materials, and their speed through rocks such as granite is approximately 3 km/sec. As *S* waves propagate, they produce a sideways shearing motion in rocks at right angles to the direction of propagation. This motion is similar to that produced in a clothesline by pulling down and letting go. Liquids (unlike rocks) are unable to spring back when subjected to sideways shear, explaining why *S* waves cannot move through liquids [9].

Surface waves cause much of the damage to buildings and other structures. Surface waves include **Love waves,** which consist of complex horizontal ground movement, and **Rayleigh waves,** with complex rolling motion (Figure 1.11). Both travel slower than body waves, but Love waves travel faster than Rayleigh waves, in general. Because different types of waves and waves of different frequency travel at different speeds away from an earthquake source area, they become organized into groups of waves traveling at similar velocities. Near the source of the large earthquake, however, there is not time for this segregation of the waves to take place, and shaking may be severe and complex. Waves traveling through rocks are both reflected and refracted across boundaries between different earth materials and at the surface of the Earth, producing amplification that may enhance shaking and damage to buildings and other structures. Furthermore, as an earthquake occurs, the propagation of waves is also affected by the rupture along a fault, which may be in a particular direction and thus tend to focus earthquake energy in that direction. Finally, surface shaking can be further complicated and accentuated by local soil conditions and topography [9].

_P_ and _S_ body waves have a wide range of frequencies, but because of rapid attenuation (loss of higher frequencies with wave propagation), most body waves tend to have frequencies of 0.5 Hz to 20 Hz. The more complex surface waves have lower frequencies (less than 1 Hz). The **frequency** of an earthquake wave equals the number of waves passing a point of reference per second, expressed as Hertz units. The **period** equals the elapsed time (in seconds) between successive peaks of a wave (observed at a point). The frequency of an earthquake wave is equal to 1 divided by the period (i.e., the reciprocal of the period). For example, the frequency of a _P_ wave with period of 2 sec is 1 divided by 2, or 0.5 Hz. Buildings and other structures commonly have natural frequencies of vibration in the same range as earthquakes. This is unfortunate because shaking of buildings is amplified when the frequency of earthquake waves is close to the natural frequency of the building. Low buildings have a higher natural frequency than taller buildings and, as a result, compressional and shear waves with relatively high frequencies tend to accentuate damage to low buildings. On the other hand, surface waves with lower frequencies tend to damage tall buildings more.

High-frequency waves attenuate (die or diminish) much more quickly with distance from a generating earthquake than do low-frequency waves. Thus, tall buildings may be damaged at relatively long distances (up to several hundred kilometers) by large earthquakes [9, 12], whereas low buildings tend to sustain the greatest damage near earthquake epicenters. This principle was dramatically illustrated in 1985 when a $M = 8.1$ earthquake several hundred kilometers away from Mexico City damaged or destroyed many of the taller buildings in the city.

Another important principle is that earth materials such as bedrock, sand and gravel, and silts and muds respond differently to seismic shaking. For example, the intensity of shaking of unconsolidated sediments may be much more severe than for bedrock. Figure 1.12 shows how the amplitude of shaking in sediments (vertical movement) is greatly increased, particularly in silt and clay deposits. This effect is called **material amplification.** A major lesson from the 1985 earthquake affecting Mexico City was that buildings constructed on materials likely to amplify seismic shaking are extremely vulnerable to earthquakes, even if the event is

FIGURE 1.12
Generalized relationship
between near-surface earth
material and amplification of
shaking during a seismic
event.

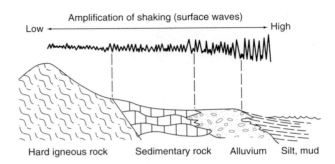

centered several hundred kilometers away. Seismic waves from this earthquake, which occurred offshore of Mexico, initially contained many different frequencies, but the seismic waves that survived the several hundred kilometer journey to the city were those with relatively long periods of 1 to 2 sec (frequencies of 1.0 Hz to 0.5 Hz). It is speculated that when these waves struck the lake beds on which Mexico City is built, the amplitude of shaking may have increased at the surface by a factor of 4 to 5 times. The intense regular shaking caused buildings to sway back and forth, and eventually many of them collapsed or "pancaked" as upper stories collapsed onto lower ones (Figure 1.13). Most of the damage was to buildings with 6 to 16 stories, because these buildings had a natural frequency that nearly matched that of the arriving seismic waves [13].

The potential for amplification of surface waves to cause damage was again demonstrated with tragic results during the 1989 $M = 7.1$ Loma Prieta (San Francisco) earthquake, when the upper tier of the Nimitz Freeway in Oakland, California, collapsed, killing 41 people (Figure 1.14). Collapse of the tiered freeway occurred on the section of roadway constructed on bay fill and mud. Where the freeway was constructed on older, stronger alluvium, less shaking occurred and the structure survived. Extensive damage was also recorded in the Marina District of San Francisco (Figure 1.15), primarily in areas constructed on bay fill and mud, including debris dumped into the bay during the cleanup following the 1906 earthquake [14].

ACTIVE FAULT ZONES

Most geologists would consider a fault to be *active* if it has moved during the past 10 k.y. (Holocene Epoch). The Quaternary Period (the past 1.65 m.y.) is the most recent period of geologic time, and most of our landscape has been produced during that time. Any fault that has moved during the Quaternary Period may be classified as *potentially active* (Table 1.3). Faults that have not moved during the past 1.65 m.y. are generally classified as *inactive*. However, it is often difficult to prove the activity of a fault in the absence of easily measured phenomena such as historical earthquakes. To prove that a fault is active, it may be necessary to determine its past earthquake history (paleoseismicity) based on the geologic record.

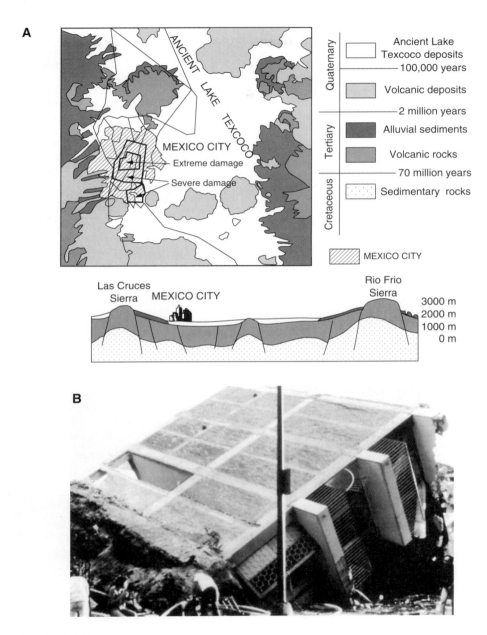

FIGURE 1.13
Earthquake damage, Mexico City, 1985. (A) Generalized geologic map of Mexico City showing ancient lake deposits where greatest damage occurred. (B) Multi-story building, one of many that collapsed.
(Map and photo courtesy of T. C. Hanks and D. Herd, U.S. Geological Survey)

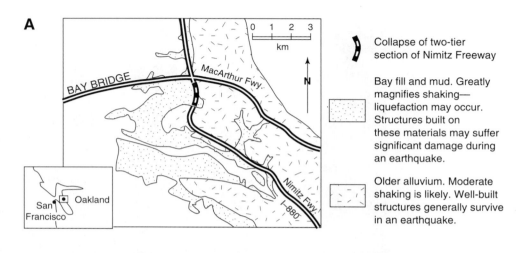

FIGURE 1.14

(A) Generalized geologic map of part of San Francisco Bay showing bay fill and mud and older alluvium. (B) Collapsed freeway.

[(A) Modified from Hough, S.E., et al., 1990. *Nature*, 344: 853–855. © Macmillan Magazines Ltd., 1990. Used by permission of the author. (B) Courtesy of John K. Nakata, U.S. Geological Survey]

FIGURE 1.15
Damage to buildings in the Marina District of San Francisco resulting from the 1989
earthquake.
(Photograph courtesy of John K. Nakata, U.S. Geological Survey)

TABLE 1.3
Terminology related to recovery of fault activity.

Geologic Age			Years Before Present	Fault Activity
Era	Period	Epoch		
Cenozoic	Quaternary	Historic (Calif.) Holocene	— 200 —	Active
			—10,000 —	
		Pleistocene		Potentially active
	Tertiary	Pre-Pleistocene	—1,650,000 —	
			— 65,000,000 —	Inactive
	Pre-Cenozoic time			
	— Age of the earth —		— 4,500,000,000 —	

(After California State Mining and Geology Board Classification, 1973)

This involves identifying faulted earth materials and determining when the most recent displacement occurred. The above definition of an active fault is used in the state of California for **seismic zoning.** However, other agencies have more conservative definitions for fault activity. For example, when considering seismic safety for nuclear power plants, the U.S. Nuclear Regulatory Commission defines a fault as "**capable**" if the fault has moved at least once in the past 50 k.y. or more than once in the past 500 k.y. These criteria provide a greater safety factor, reflecting increased concern for the risk of siting nuclear power plants.

SLIP RATES AND RECURRENCE INTERVALS

Our discussion of faults and earthquakes involves two important concepts: slip rates on faults, and recurrence intervals, or repeat times, of earthquakes. **Slip rate** on a fault is defined as the ratio of slip (displacement) to the time interval over which that slip occurred. For example, if a fault has moved 1 m during a time interval of 1000 yr, the slip rate is 1 mm/yr. The **average recurrence interval** on a particular fault is defined as the average time interval between earthquakes, and it may be determined by three methods:

1. Paleoseismic data: Averaging the time intervals between earthquakes recorded in the geologic record (see Chapter 8).
2. Slip rate: Assuming a given displacement per event and dividing that number by the slip rate. For example, if the average displacement per event is 1 m and the slip rate is 2 mm/yr, then the average recurrence interval would be 500 yr.
3. Seismicity: Using historical earthquakes and averaging the time intervals between events.

Defining the terms *slip rate* and *recurrence interval* is easy, and the calculations seem straightforward, but the underlying concepts are far from simple. Fault slip rates and recurrence interval tend to be variable in time, casting suspicion on average rates that might be derived. For example, it is not uncommon for earthquake events to be clustered in time and then be separated by relatively long periods of low activity. Both slip rate and recurrence interval will vary depending upon the time interval for which data is available. The topics of slip rates and recurrence intervals will be discussed repeatedly in this book. They are introduced here to facilitate later discussions.

ESTIMATION OF SEISMIC RISK

Catastrophic earthquakes are devastating events. Historic earthquakes have destroyed large cities and taken thousands of lives in a matter of seconds. Table 1.4 lists some of the major historical earthquakes that have occurred in the United States.

TABLE 1.4
Selected major earthquakes in the United States.

Year	Locality	Damage $ Million	Lives Lost
1811–12	New Madrid, Missouri	Unknown	
1886	Charleston, South Carolina	23	60
1906	San Francisco, California	524	700
1925	Santa Barbara, California	8	13
1933	Long Beach, California	40	115
1940	Imperial Valley, California	6	9
1952	Kern County, California	60	14
1959	Hebgen Lake, Montana (damage to timber and roads)	11	28
1964	Alaska and U.S. West Coast (includes tsunami damage from earthquake near Anchorage)	500	131
1965	Puget Sound, Washington	13	7
1971	San Fernando, California	553	65
1983	Coalinga, California	31	—
1983	Central Idaho	15	2
1987	Whittier, California	358	8
1989	Loma Prieta (San Francisco), California	5,000	62
1992	Landers, California	27	1
1994	Northridge, California	>20,000	61

(Modified after Hays, 1981 [12])

Seismic risk maps have been prepared for the United States. Although relative hazard (where earthquakes have occurred) has been used to construct these maps, calculating the probability of a particular event or amount of shaking is considered preferable. Figure 1.16 is an earthquake hazard map, showing contours of maximum horizontal ground acceleration caused by seismic shaking that are not likely (with a 90% probability) to be exceeded in 50 years. One way of interpreting the map is that the darkest areas on Figure 1.16 represent the regions of greatest seismic hazard, because those areas are most likely to experience the greatest seismic shaking (in this case, horizontal ground acceleration) in an average 50-yr interval. Although regional earthquake hazard maps are valuable, considerably more data are necessary to evaluate hazardous areas more precisely in order to develop building codes and determine insurance rates.

In California, **conditional probabilities** (probability dependent upon known or estimated conditions) of major earthquakes along segments of the San

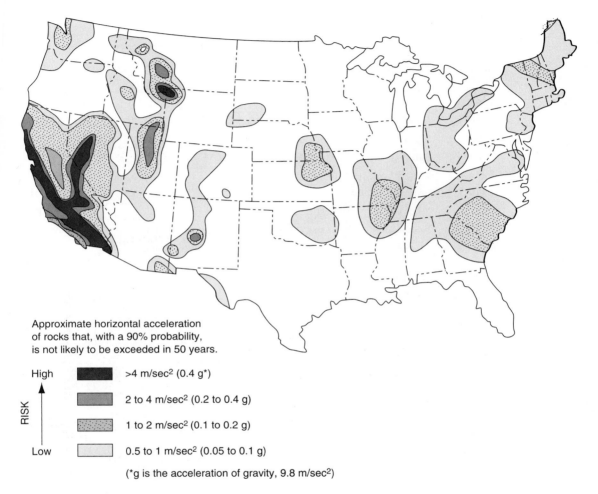

Approximate horizontal acceleration
of rocks that, with a 90% probability,
is not likely to be exceeded in 50 years.

High   ████   >4 m/sec² (0.4 g*)

            ████   2 to 4 m/sec² (0.2 to 0.4 g)

RISK   ████   1 to 2 m/sec² (0.1 to 0.2 g)

Low   ████   0.5 to 1 m/sec² (0.05 to 0.1 g)

            (*g is the acceleration of gravity, 9.8 m/sec²)

FIGURE 1.16
A probabilistic approach to the seismic hazard in the United States. The darker the area
on the map, the greater the hazard.
(From Algermissen and Perkins, 1976. *U.S. Geological Survey Open File Report* 76-416)

Andreas fault for a 30-yr period have been calculated (Figure 1.17). The probabilities were calculated following synthesis of historical records and geologic
evaluation of prehistoric earthquakes [15]. In 1988, this approach assigned a
probability of about 30% for a major event on the San Andreas fault segment
through the Santa Cruz Mountains where the M = 7.1 Loma Prieta earthquake
occurred on October 17, 1989. Occurrence of this earthquake supports the validity of the conditional-probability approach. The probability of a large earthquake
on the southern segment of the San Andreas fault is estimated to be close to

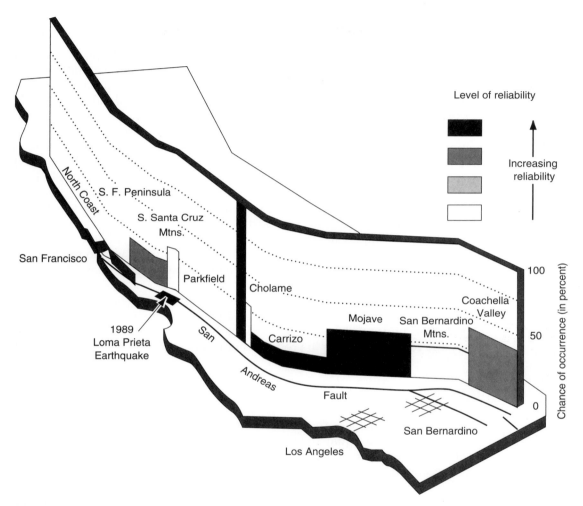

FIGURE 1.17
Conditional probability of a major earthquake along segments of the San Andreas fault
(1988–2018).
(After Heaton et al., 1989 [15])

50% for the next 30 years. The $M = 7.5$ Landers earthquake that occurred east of
the San Andreas fault in 1992 was a surprising event. That event produced
major right-lateral horizontal surface displacement of up to 5 m, and maximum
Modified Mercalli intensity of VIII [16] on a fault system that was previously
mapped but which had not received much attention. This large earthquake
caused relatively little damage ($27 million) and one death, primarily because it
occurred in a region with low buildings and few people.

EFFECTS OF EARTHQUAKES

Primary effects of earthquakes are caused directly by the earthquake and can include violent ground-shaking motion accompanied by surface rupture and permanent displacement. For example, the 1906 earthquake at San Francisco produced 6.5 m of horizontal displacement and a maximum Modified Mercalli intensity of XI [9]. Such violent motions can produce surface accelerations that snap and uproot large trees and knock people to the ground. This motion may shear or collapse large buildings, bridges, dams, tunnels, and pipelines, as well as other rigid structures [17]. The great 1964 Alaskan earthquake (M_L = 8.25) caused extensive damage to railroads, airports, and buildings. The 1989 Loma Prieta (San Francisco) earthquake, with a M = 7.1, was much smaller than the Alaska event and yet caused about $5 billion in damage. The 1994 Northridge earthquake, with M = 6.7, was one of the most expensive disasters ever in the United States, causing 61 deaths and inflicting more than $20 billion in damage. The Northridge event caused so much damage because there was so much there to be damaged—the Los Angeles region is highly urbanized and has a high population density.

Short-term secondary effects of earthquakes include liquefaction, landslides, fires, seismic seawaves (tsunami), and floods (following collapse of dams). Long-term secondary effects include regional subsidence or emergence of landmasses and regional changes in groundwater levels.

Liquefaction

Liquefaction is defined as the transformation of water-saturated granular material from a solid to a liquid state. During earthquakes, this may result from an increase in pore-water pressure caused by compaction during intense shaking. Liquefaction of near-surface water-saturated silts and sand causes the materials to lose their shear strength and flow. As a result, buildings may tilt or sink into the liquefied sediments, and tanks or pipelines buried in the ground may float to the surface [18].

Landslides

Earthquake shaking commonly triggers many landslides (a comprehensive term for several types of hillslope failure) in hilly and mountainous areas. Landslides can be extremely destructive and cause great loss of life, such as during the 1970 Peru earthquake. In that event, more than 70,000 people died; of this total, 20,000 were killed by a giant landslide that buried several towns. Both the 1964 Alaskan earthquake and the 1989 Loma Prieta earthquake caused extensive landslide damage to buildings, roads, and other structures.

Fires

Fire is a major secondary hazard associated with earthquakes. Shaking of the ground and surface displacements can break electrical power and gas lines and ignite fires. In individual homes and other buildings, appliances such as gas heaters may be knocked over. The threat from fire is doubled because firefighting equipment may be damaged and water mains may be broken. Earthquakes in both Japan and the United States have been accompanied by devastating fires. The San Francisco earthquake of 1906 has been called the "San Francisco Fire"; in fact, 80% of the damage from that event was caused by firestorm that ravaged the city for several days. The 1989 Loma Prieta earthquake also caused large fires in the city's Marina District. Perhaps the most lethal earthquake-induced fire occurred in 1923 in Japan. The earthquake killed 143,000 people, and 40% of them died in a firestorm that engulfed an open space where people had gathered in an unsuccessful attempt to reach safety [17]. The 1995 Kobe, Japan, earthquake ruptured gas lines, and fires devastated parts of the city. Ruptured water lines and damaged roads prevented firefighters from reaching and extinguishing fires.

Tsunami

Tsunami, or seismic seawaves, can be extremely destructive and present a serious natural hazard (Figure 1.18). Fortunately, damaging tsunami are relatively rare and usually are confined to the Pacific Basin. The frequency of these events in the United States is about one every eight years [19]. Tsunami originate when ocean water is displaced vertically during large earthquakes, submarine mass movements, or submarine volcanic eruptions. In open water, the waves may travel at speeds as great as 800 km/hr, and the distance between successive crests may exceed 100 km. Wave heights in deep water may be less than 1 m, but when the waves enter shallow coastal waters they slow to less than 60 km/hr and wave heights may increase to 15 m or more. Tsunami claimed most of the lives lost in the 1964 Alaskan earthquake. Tsunami can cause catastrophic damage thousands of kilometers from where they are generated. For example, in 1960, an earthquake originating in Chile caused a tsunami that reached Hawaii, killing 61 people. Tsunami often can be detected in time to warn coastal communities that lie in the tsunami's path. Travel time for a tsunami from Chile to Hawaii is about 15 hours, and from the Aleutian Islands in Alaska to northern California the time is about four hours. The hazard from a tsunami at a particular coastal site depends in part on local coastal and sea-floor topography that may increase or decrease wave height [9]. Following an earthquake that produces a tsunami, the arrival time of the seismic seawaves can often be estimated to within plus or minus 1.5 min/hr of travel time. This information has been used to produce a tsunami warning system such as that shown for Hawaii in Figure 1.19. Consideration is now being given to produce other tsunami warning

FIGURE 1.18
Tsunami damage to fishing boats at Kodiak, Alaska, caused by the 1964 earthquake.
(Photograph courtesy of NOAA)

systems, to warn people in northwestern California of tsunami generated by Alaskan or Cascadia subduction-zone earthquakes. The plan involves placing instruments on the bottom of the Pacific Ocean (four off the Alaskan coast and three off the northwest coast of California) that would detect and transmit data on the movement of tsunami to warn coastal cities in their path.

Damage caused by tsunami is most severe at the water's edge, where boats, harbors, buildings, transportation systems, and utilities may be destroyed. The waves also may cause damage to aquatic and supratidal life in the nearshore and onshore environments [19].

Waves caused by landslides also can cause damage. In 1958, an earthquake induced a landslide into Lituya Bay, Alaska, causing a truly giant wave that produced run-up on land to an elevation of over 500 m above sea level [20].

Regional Changes in Land Elevation

Vertical deformation, including both uplift and subsidence, is another secondary effect of some large earthquakes. The great ($M = 8.25$) 1964 Alaskan earthquake, with Modified Mercalli intensity of X–XI [9] caused vertical deformation over an

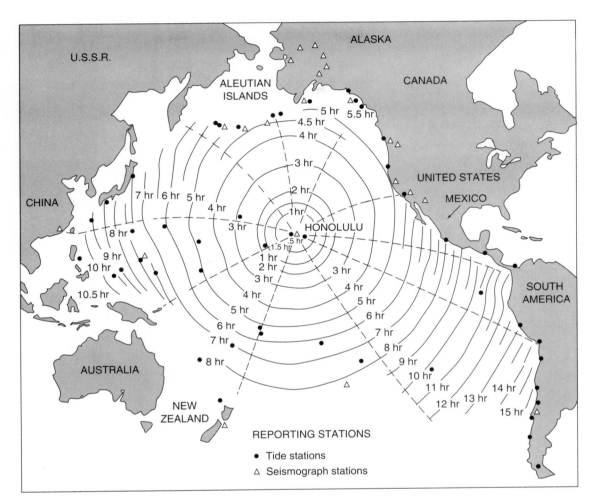

FIGURE 1.19
Tsunami warning system. Map shows reporting stations and tsunami travel times to
Honolulu, Hawaii.
(From NOAA)

area of more than 250,000 km² [21]. The deformation included two major zones
of warping, each about 500 km long and more than 210 km wide (Figure 1.20),
including uplift as much as 10 m and subsidence as much as 2.4 m. These regional
changes in land level caused effects ranging from severely disturbing coastal
marine life to changes in groundwater levels. As a result of subsidence, flooding
occurred in some communities, whereas in areas of uplift, canneries and fishermen's homes were displaced above the high-tide line, rendering docks and other
facilities inoperable. In 1992, a major earthquake ($M = 7.1$) near Cape Mendocino

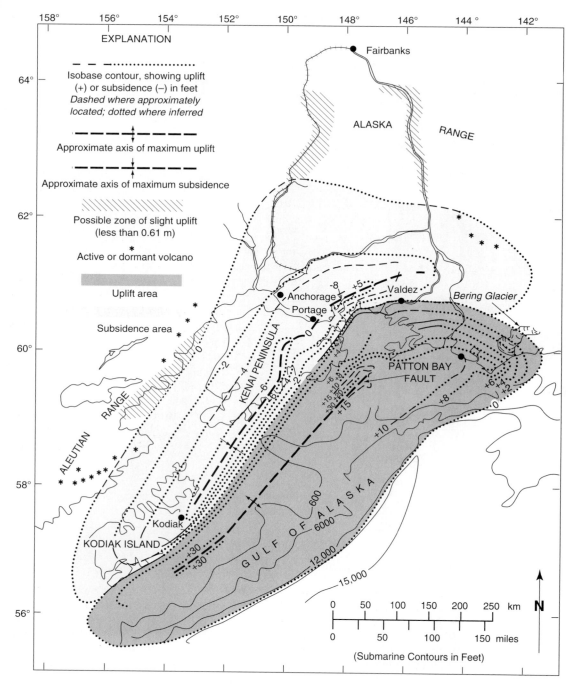

FIGURE 1.20

Map showing the distribution of tectonic uplift and subsidence in south-central Alaska caused by the Alaskan earthquake of 1964 (2 to 30 ft = 0.6 to 9.2 m).

(From Eckel, 1970. *U.S. Geological Survey Professional Paper* 5460)

in northwestern California produced approximately 1 m of uplift at the shoreline, resulting in the deaths of communities of marine organisms exposed by the uplift [22] (see Chapter 6).

TECTONIC CREEP

Tectonic creep is the process of displacement along a fault zone that is not accompanied by perceptible earthquakes. The process can slowly damage roads, sidewalks, building foundations, and other structures. Tectonic creep has damaged culverts under the football stadium of the University of California at Berkeley, and periodic repairs have been necessary. Movement of approximately 3.2 cm in 11 years was measured (Figure 1.21) [23]. More rapid rates of tectonic creep have been recorded on the Calaveras fault zone, a segment of the San Andreas fault near Hollister, California. At one location, a winery located on the fault is slowly being pulled apart at about 1 cm/yr [24]. Damages resulting from tectonic creep generally occur along narrow fault zones subject to slow, continuous displacement. However, creep may also be discontinuous and variable in rate.

EARTHQUAKES CAUSED BY HUMAN ACTIVITY

Several human activities are known to cause earthquakes or to increase earthquake activity. Damage from these earthquakes is regrettable, but the lessons learned may help control or stop large catastrophic earthquakes in the future. Three ways that the actions of people have caused earthquakes are [25]:

- Loading the Earth's crust, such as building a dam and reservoir (reservoir-induced seismicity)
- Injecting liquid waste deep into the ground through disposal wells
- Underground nuclear explosions

During the first ten years following the completion of the Hoover Dam on the Colorado River in Arizona and Nevada, several hundred local tremors occurred. Most of these were very small, but one had a magnitude of about 5, and two had magnitudes of about 4 [25]. An earthquake—attributed to reservoir-induced seismicity—of magnitude about 6 in India killed about 200 people following dam construction and filling of a reservoir. Evidently, faults may be activated by the increased load of water on the land and by increased water pressure in the rocks below the reservoir.

From April 1962 to November 1965, several hundred earthquakes occurred in the Denver, Colorado, area. The largest earthquake had a $M = 4.3$ and knocked bottles off store shelves. The source of the earthquakes was eventually traced to the Rocky Mountain Arsenal, which was manufacturing materials for chemical

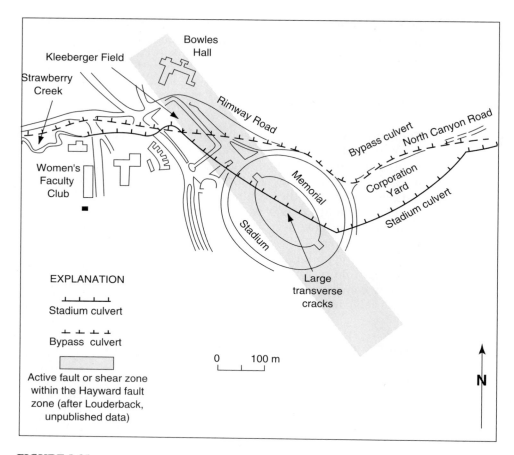

FIGURE 1.21
Map showing the location of the University of California at Berkeley Memorial Stadium,
the active fault or shear zone within the Hayward fault zone, and the stadium culvert
where major cracking has taken place
(After Radbruch et al., 1966. *U.S. Geological Survey Circular* 525)

warfare. Liquid waste from the manufacturing process was being pumped down a
disposal well to a depth of about 3600 m. The rock receiving the waste was a
highly fractured metamorphic unit, and injection of liquid increased the fluid
pressure, facilitating slippage along preexisting fractures and producing the earthquakes. Study of the earthquake activity revealed a strong correlation between
the rate of injection of waste and the occurrence of earthquakes. When waste
injection stopped, the earthquakes stopped [26]. These induced earthquakes in
the Denver area were a milestone because they alerted scientists to the fact that
earthquakes and fluid pressure are related.

Numerous earthquakes with magnitudes as large as 5.0 to 6.3 have been
triggered by underground explosions at the U.S. nuclear test site in Nevada [25].

Analysis of the aftershocks suggests that the explosions caused some release of natural tectonic strain. This led to discussions by scientists as to whether nuclear explosions might be used to prevent large earthquakes by releasing strain before it reached a critical point and caused a large earthquake. These discussions never resulted in serious consideration of actual application.

THE EARTHQUAKE CYCLE

Observations of the 1906 San Francisco earthquake led to a model known as the **Earthquake Cycle.** Important features of the hypothesis are related to drop in elastic strain following an earthquake and reaccumulation of strain prior to the next event. **Strain** was defined previously as deformation (displacement or change in shape or volume) resulting from stress, and **elastic strain** may be thought of as deformation that is not permanent, provided that the stress is released. If the strain is released, the deformed material returns to its original shape; for example, when a rubber band is stretched and released or when an archery bow is bent and released. During an earthquake, elastic strain drops because there is a stress drop when the rocks break and permanent displacement occurs (the rubber band or bow breaks). This process is referred to as **elastic rebound** and is illustrated on Figure 1.22. It takes time for sufficient elastic strain to accumulate again to produce another earthquake [11]. The Earthquake Cycle is discussed further in Chapter 3.

Several models have been proposed as part of the Earthquake Cycle to describe earthquakes and slip that recur on fault zones [27, 28]. If a particular fault or fault segment tends to generate earthquakes with about the same maximum magnitude, then that fault is said to generate **characteristic earthquakes.** The **characteristic earthquake model** resulted from paleoseismological studies of the Wasatch fault zone near Salt Lake City, Utah, and the San Andreas fault zone, California. In the characteristic earthquake model: (1) the displacement per event at a point on the fault is constant, (2) the slip rate along the length of the fault or fault segment may be variable, and (3) the size of large earthquakes is nearly constant (the range in magnitude for an event is narrow and near the maximum, and moderate events are infrequent) [27]. A competing concept is the **uniform slip model,** also developed for the San Andreas fault zone [28]. In the uniform slip model there is: (1) constant displacement per event at a point along the fault, (2) constant slip rate along the length of the fault or fault segment, and (3) a constant size of earthquakes (more frequent moderate earthquakes may occur). One difference between the characteristic earthquake and uniform slip models is that the former may have a variable slip rate along the length of the fault. If the slip that results from a particular earthquake is known to vary along the fault, then the long-term slip rate must also be variable [29]. Since displacement dies out at ends of a rupture, displacement is variable. The main difference, however, is that the characteristic earthquake model predicts the recurrence of large earthquakes with infrequent, moderate events, whereas the uniform slip model allows for more frequent

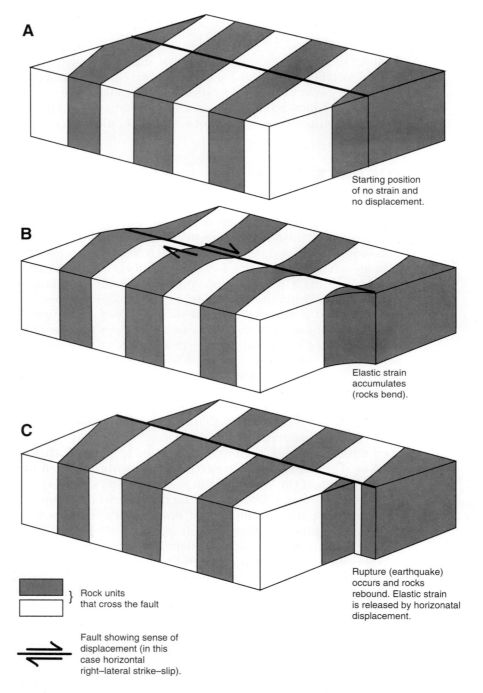

A Starting position of no strain and no displacement.

B Elastic strain accumulates (rocks bend).

C Rupture (earthquake) occurs and rocks rebound. Elastic strain is released by horizonatal displacement.

} Rock units that cross the fault

Fault showing sense of displacement (in this case horizontal right–lateral strike–slip).

FIGURE 1.22
Idealized block diagrams illustrating the earthquake cycle and elastic rebound. (A) Beginning position with no strain or displacement. (B) After accumulation of elastic strain. (C) Following earthquake and rupture. (Courtesy of Fred Duennebier)

moderate events. A third model is known as the **variable slip model,** in which the amount of slip and the length of rupture may both vary from one earthquake to the next, producing variability in earthquake size [27, 29].

As yet, we do not have sufficient paleoseismic data on individual faults to determine which of the models (characteristic earthquake, uniform slip, or variable slip) best characterizes recurrent earthquakes and slip on faults. In fact, all three models are apparently present in nature. The characteristic earthquake model seems to fit for the Wasatch fault and perhaps for some segments of the San Andreas fault [27]. The uniform slip model may also be invoked for some segments of the San Andreas fault [28]. Still other faults are likely to have displacement histories and earthquakes that are best represented by a variable slip model. Variable slip need not be a random pattern of earthquake size and displacement, as sometimes suggested [27]. Rather, it is hypothesized (assumed without proof) that the more we learn about the behavior of particular faults (especially reverse faults; see Figure 1.7), the more variable (but perhaps *systematically* variable) the pattern of earthquakes is likely to be. For example, three subduction-zone earthquakes (reverse faulting) in 1707, 1854, and 1946, that uplifted the coastal area near Kyoto, Japan (Nankaido earthquakes), produced rupture lengths that varied from 300 km to 500 km and uplift that varied from 1.2 m to 1.8 m. Similarly, a series of earthquakes off the Colombia-Ecuador coast in 1906, 1942, 1958, and 1979 ruptured the same 500-km length of subduction zone; the first event had a seismic moment over five times greater than the moments of the other three events, suggesting the fault slip in the 1906 event was much greater than the other three [29]. The Japan and Colombia-Ecuador events support the variable slip model.

Some faults may be characterized by clusters of earthquakes over periods of a 1000 yr or so, followed by long, quiet periods before new clusters of events occur. This pattern of activity may fit the Oued Fodda fault which produced the 1980 Algeria earthquake ($M = 7.3$). Paleoseismic investigation suggests that the three most recent earthquakes occurred during the last 900 yr (including the 1980 event) and provides a recurrence interval of about 450 yr. However, these recent earthquakes represent a clustering of events large enough to produce surface rupture. During the Late Pleistocene, the earthquake activity on the fault was characterized by relatively short periods, with earthquakes occurring every few hundred years, separated by much longer periods (thousands to tens of thousands of years) with no large earthquakes [30].

Although we have many empirical observations concerning physical changes in earth materials before, during, and after earthquakes, there is no general agreement on a physical model to explain the observations. One model, known as the **dilatancy diffusion model** [9, 31], assumes that the first stage in earthquake development is an increase of elastic strain in rocks that causes them to **dilate,** or undergo an inelastic increase in volume after the stress on the rock reaches one-half its breaking strength. During dilation, open fractures develop in the rocks, and at this stage the first physical changes take place that might indicate a future earthquake. The model assumes that the dilatancy and fracturing of the rocks are first associated with a relatively low water pressure in the dilated rocks (stage 2, Figure 1.23), which helps to produce lower seismic velocity, more earth movement, higher radon gas emission (radon is a naturally-radioactive gas that is dissolved in water and released as rocks fracture and dilate), lower electrical resistivity, and fewer minor seismic events. Water then enters the open fractures (stage 3, Figure 1.23), causing the pore pressure to increase (which increases the seismic velocity while further lowering electrical resistivity), thus weakening the rocks and triggering an earthquake (stage 4). After the movement and release of stress, the rocks resume many of their original characteristics (stage 5) [31].

There is considerable controversy concerning the validity of the dilatancy diffusion model. One aspect of the model gaining considerable favor is the role of **fluid pressure** (force per unit area exerted by a fluid) in earthquakes. As we learn more about rocks at seismogenic depths (the depth where earthquakes originate), it is apparent that a lot of water is present. Deformation of the rocks and a variety of other processes are thought to increase the fluid pressure at depth, and this lowers the shear strength. If the fluid pressure becomes sufficiently high, then this can facilitate earthquakes. A wide variety of data from several environments including subduction zones and active fold belts suggest that high fluid pressures are present in many areas where earthquakes occur. Thus, there is increasing speculation and interest in the role of fluid flow that affects fault displacement and is intimately related to the earthquake cycle. This process has been termed the **fault-valve mechanism** [32]. The mechanism is a hypothesis in which fluid pressure rises until failure occurs, thus triggering an earthquake and discharging fluid upward. Subsequent sealing of the rock matrix in the fault zone allows fluid pressure to reaccumulate, initiating another cycle.

PREDICTING GROUND MOTION

Engineering design of critical facilities such as power plants and dams requires careful evaluation of earthquake hazard. Of particular importance is prediction of **strong ground motion** due to earthquakes that may occur at or near facility sites. Seismographs provide information about the amplitude of seismic shaking, as illustrated on Figure 1.10A. Instruments known as **accelographs** measure and record vertical and horizontal accelerations produced by earthquakes. By measuring the vertical and horizontal components of acceleration in both the north-south and east-west directions, a three-dimensional picture of ground acceleration is created. Another important parameter is the **duration of shaking.** For ground accelerations measured from an accelograph, the duration of strong shaking is defined as the **bracketed duration,** which is the time in which the acceleration is at least 0.05g. For the example shown on Figure 1.24, the duration of strong shaking is approximately 8 sec. Thus we recognize that the **peak maximum acceleration** of seismic waves is important, but duration of strong shaking is also important. [9].

FIGURE 1.23
Dilatancy diffusion model to explain the mechanism responsible for triggering earthquakes. The curves show the expected precursory signals.
(After Press, 1975. *Earthquake Prediction.* © May 1975 by Scientific American, Inc. All rights reserved.)

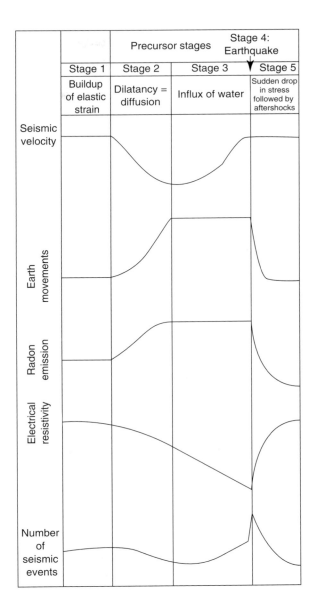

Assessment of earthquake hazard starts with identification of the **tectonic framework** (geometry and spatial pattern of faults or seismic sources) in order to predict earthquake ground motion (Figure 1.25). The major objective of such assessment is to develop **time histories** (relationships between properties of seismic waves and time) of ground motion resulting from the largest earthquakes that could shake the site of interest. The process of predicting ground

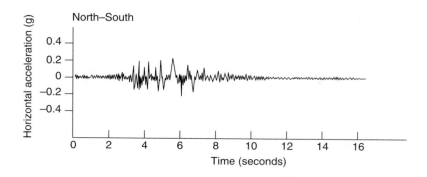

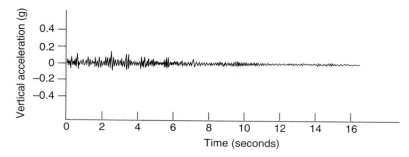

FIGURE 1.24

Hypothetical graph of vertical and horizontal accelerations from an earthquake with a magnitude $M_w = 6.5$ at a distance of about 40 km from the center of energy release. Time "0" on the graph is the first arrival of the P waves. Vertical accelerations in this example are approximately 0.1g. On the graph which shows the north-south horizontal acceleration, the S and L waves arrive approximately 4 sec later than P waves with a maximum acceleration of approximately 0.25g.

motion from a given earthquake may be illustrated by considering a hypothetical example. Figure 1.26A shows a map of a dam and reservoir site. The objective is to predict strong ground motion at the dam from several seismic sources (faults) in the area. The tectonic framework shown consists of a north-dipping reverse fault and an associated fold (an anticline) located to the north of the dam, as well as a right-lateral strike-slip fault located to the south of the dam. Figure 1.26B shows a cross section through the dam illustrating the geologic environment, including several different earth materials, folds, and faults. Assuming that earthquakes would occur at depths of approximately 10 km, the distances from the dam to the two seismic sources (the reverse fault and the strike-slip fault) are 42 km and 32 km, respectively. Thus, for this area, two focal mechanisms are possible: reverse faulting and strike-slip faulting. The next step in the process is to estimate the largest earthquakes likely to occur on these faults. Assume that field work in the area revealed ground rupture and other evidence of faulting in the past, suggesting that, on the strike-slip fault, approxi-

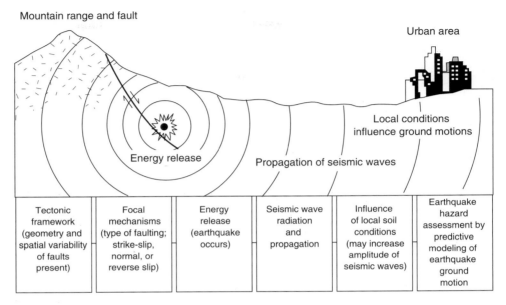

Mountain range and fault

Urban area

Energy release

Local conditions
influence ground motions

Propagation of seismic waves

| Tectonic framework (geometry and spatial variability of faults present) | Focal mechanisms (type of faulting; strike-slip, normal, or reverse slip) | Energy release (earthquake occurs) | Seismic wave radiation and propagation | Influence of local soil conditions (may increase amplitude of seismic waves) | Earthquake hazard assessment by predictive modeling of earthquake ground motion |

FIGURE 1.25
Assessment of earthquake hazard by modeling of ground motion.
(After Vogel, A., 1988. Earthquake prognostics—the development of a concept for interdisciplinary work. In A. Vogel and K. Brandes (eds.), *Earthquake Prognostics.* Friedr. Vieweg & Sohn: Brannschweig/Weisbaden: 1–13)

mately 50 km of fault length might rupture in a single event, with right-lateral strike-slip motion of 2 m. The field work also revealed that the largest rupture likely on the reverse fault would be 30 km of fault length, with vertical displacement of about 1 m. Given this information, the magnitudes of possible earthquake events can be estimated from graphs such as those shown on Figure 1.27 [33, 34]. Fifty kilometers of surface rupture are associated with an earthquake of approximately $M = 7$. Similarly, for the reverse fault with surface rupture length of 30 km, the magnitude of a possible earthquake is estimated to be 6.5. Notice on Figure 1.27 that the regression line that predicts the moment magnitude is for strike-slip, normal, and reverse faults. Statistical analyses have suggested that the relation between moment magnitude and length of surface rupture is not sensitive to the style of faulting [34].

With the information above, the next task is to estimate the seismic shaking or ground motion expected from these events. These are referred to as **response spectra,** which are relationships between ground motion and period of earthquake waves [35]. There are two approaches available to estimate response spectra:

- Empirical evaluation where ground motions have been recorded from previous events. Fortunately, there are a large number of records from earthquakes around the world where seismographs and accelographs

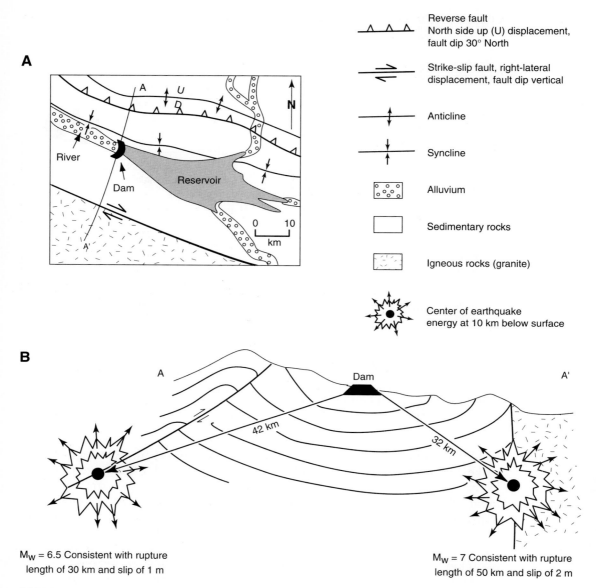

A

Reverse fault
North side up (U) displacement,
fault dip 30° North

Strike-slip fault, right-lateral
displacement, fault dip vertical

Anticline

Syncline

Alluvium

Sedimentary rocks

Igneous rocks (granite)

Center of earthquake
energy at 10 km below surface

River

Dam

Reservoir

0   10
km

N

B

A

Dam

A'

42 km

32 km

M_W = 6.5 Consistent with rupture
length of 30 km and slip of 1 m

M_W = 7 Consistent with rupture
length of 50 km and slip of 2 m

FIGURE 1.26
Tectonic framework for a hypothetical dam site. (A) Geologic map. (B) Cross section
A–A', showing seismic sources and distances of possible ruptures to the dam.

have recorded seismic shaking. This technique involves identifying those
records that most closely approach the conditions for the site of interest.
The objective is to match as closely as possible the tectonic framework,
rock types, type of faulting, and earthquake magnitude from a known
event to known conditions at the dam site. The assumption is that the
shaking and strong ground motion associated with a known earthquake

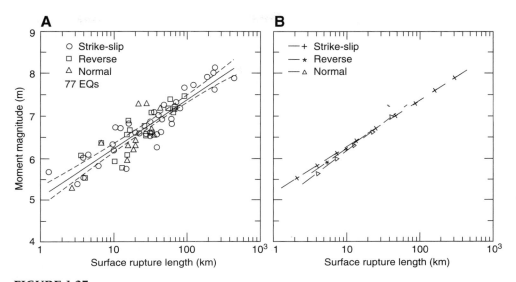

FIGURE 1.27

Relationship between moment magnitude of an earthquake and surface rupture length. Data for 77 events are shown in (A). Solid line is the "best-fit" or regression line, and dashed lines are error bars at the 95% confidence level. Graph in (B) shows individual lines for different types of faults. There is no significant difference between the lines. (After Wells and Coppersmith, 1994 [34])

will produce similar ground motion at the dam. For example, an earthquake of $M = 6.4$ on a reverse fault in Afghanistan might serve as a model for the dam site used as an example here. Similarly, an earthquake of $M = 7.1$ on the San Andreas fault in California may be used as a model to estimate the ground motion from such an event on the strike-slip fault shown on Figure 1.26. Of course, it is difficult to exactly match the conditions from a known event to those of the dam site, and so allowances are necessary to adjust for small differences in earthquake magnitude and distances to the predicted epicenters. If the match between a known event and possible event at the dam site is fairly good, then ground-motion parameters such as duration of shaking and average peak acceleration of ground motion may be estimated.

• The second approach is to develop theoretical or numerical models to estimate ground motions, including acceleration time histories for the various faulting scenarios [35, 36]. Such models are now commonly used in the seismic-risk evaluation of structures such as dams, bridges, and tall buildings.

Results from the modeling of ground motion are then compared to the empirical results and, if the agreement is good, the ground-motion parameters can be used to evaluate potential shaking, duration of shaking, and ground acceleration at the dam site. If this analysis is completed prior to the designing of a dam, then the parameters are useful to engineers designing the dam to minimize potential damage from earthquakes. If the evaluation is of an existing dam site, then the information is useful in evaluating whether additional engineering work is necessary to upgrade the structure to render it more resistant to earthquakes.

The preceding discussion of methods used to assess earthquake hazard outlines procedures for evaluating the seismic hazard at power plants and dams. Although this methodology suffers from limiting assumptions and shortcomings, it certainly is a valuable tool insofar as it allows estimation of strong ground motions likely to affect a particular site. As additional earthquake records are obtained, and in particular strong ground motions close to the epicentral areas are recorded, then our understanding of how to better design structures to withstand earthquake shaking will improve.

SUMMARY

Tectonics refers to processes and landforms resulting from deformation of the Earth's crust, and active tectonics refers to those tectonic processes that produce deformation of the Earth's crust on a time scale that is significant to humans. Active tectonics includes slow disruption of the crust that may cause damage to human structures, but is most concerned with catastrophic events such as earthquakes that cause severe damage to people, property, and society. Tectonic processes are driven by forces deep within the Earth, and collectively these processes are known as the tectonic cycle. Global tectonics, or plate tectonics, have been studied intensively and have gained the status of a unified theory of how the Earth works.

A fault is a fracture or fracture system along which rocks have been displaced. A group of related faults or fault traces is known as a fault zone, and most major fault zones are segmented. Fault segments are recognized on the basis of changes in the fault-zone morphology or geometry as well as seismic and paleoseismic activity.

Both magnitude and intensity of earthquakes are important in evaluating potential earthquake hazard. Although the Richter magnitude has been used for many years, it is now being replaced by the more physically based moment magnitude system. The Modified Mercalli Scale is based on observations concerning severity of shaking and response of structures to earthquakes. It is a valuable tool in assessing earthquake hazards.

Active fault zones are those for which it can be demonstrated that a fault has moved during the past 10 k.y. Faults that have moved in the past 2 m.y. are considered potentially active, and those that have not moved during that period generally are classified as inactive. Other, more conservative classifications for fault activity also exist. For example, the U.S. Nuclear Regulatory

Commission considers a fault to be "capable" if it has moved once in the past 50 k.y., or more than once in the last 500 k.y. Estimation of seismic risk is an important endeavor and involves development of seismic risk maps and calculation of conditional probabilities of earthquakes occurring in the future. Slip rate on a fault is an important parameter in estimating the earthquake hazard for a particular fault. The average recurrence interval of earthquakes on a particular fault also is an important characteristic and can be determined by: paleoseismic data, the ratio of assumed displacement per event to the slip rate, and historical seismicity.

Effects of earthquakes include violent shaking, surface rupture, liquefaction, landslides, fires, tsunamis, and regional changes in land elevation. Earthquakes also have been caused by human activity, such as loading the crust by building a reservoir or injecting liquid waste deep in the Earth in disposal wells. An important concept is the Earthquake Cycle, which is related to drop in elastic strain following an earthquake and reaccumulation of strain prior to the next event. Several models have been proposed as part of the Earthquake Cycle, including characteristic earthquake, uniform slip, and variable slip models.

An important aspect of earthquake-hazard reduction is prediction of strong ground motion. Assessment of the earthquake hazard starts with identification of the tectonic framework (geometry and spatial pattern of faults or seismic sources) followed by identification of possible focal mechanisms for faults present in the area being evaluated. Field work and other evidence are used to predict the magnitude of earthquake events that might be expected at a particular site. This information is used with mathematical models and in comparisons with known earthquakes to predict strong ground motion at a site of interest. This information then may be applied to better designed structures such as dams, power plants, and freeway overpasses.

REFERENCES CITED

1. Geophysics Study Committee, 1986. Overview and recommendations. In *Active Tectonics*. National Academy Press: Washington, D.C.

2. Davis, G.H., 1993. Basic science planning in "active tectonics." *EOS: Transactions, American Geophysical Union*, 74 (43): 59.

3. White, G.F., and J.E. Haas, 1975. *Assessment of research on natural hazards*. The MIT Press: Cambridge, MA.

4. Advisory Committee on the International Decade for Natural Hazard Reduction, 1989. *Reducing disaster's toll*. National Academy Press: Washington, D.C.

5. Le Pichon, X., 1968. Sea-floor spreading and continental drift. *Journal of Geophysical Research*, 73: 3661–3697.

6. Isacks, B., J. Oliver, and L. Sykes, 1968. Seismology and the new global tectonics. *Journal of Geophysical Research*, 73: 5855–5899.

7. Dewey, J.F., 1972. Plate tectonics. *Scientific American*, 22: 56–68.

8. Hamilton, R.M., 1980. Quakes along the Mississippi. *Natural History*, 89: 70–75.

9. Bolt, B.A., 1988. *Earthquakes*. W.H. Freeman: San Francisco.

10. Hart, E.W., W.A. Bryant, and J.A. Treiman, 1993. Surface faulting associated with the June 1992 Landers earthquake, California. *California Geology*: 10–16.

11. Hanks, T.C., 1985. The national earthquake hazards reduction program: scientific status. *U.S. Geological Survey Bulletin* 1659.

12. Hays, W.W., 1981. Facing geologic and hydrologic hazards. *U.S. Geological Survey Professional Paper* 1240-B.

13. Jones, R.A., 1986. New lessons from quake in Mexico. *Los Angeles Times,* September 26.

14. Hough, S.E., P.A. Friberg, R. Busby, E.F. Field, K.H. Jacob, and R.D. Borcherdt, 1989. Did mud cause freeway collapse? *EOS: Transactions, American Geophysical Union,* 70 (47): 1497–1504.

15. Heaton, T.H., D.L. Anderson, W.J. Arabasz, R. Buland, W.L. Ellsworth, S.H. Hartzell, T. Lay, and P. Spudich, 1989. National seismic system science plan. *U.S. Geological Survey Circular* 1031.

16. Toppozada, T.R., 1993. The Landers–Big Bear earthquake sequence and its felt effects. *California Geology,* January-February: 3–9.

17. Office of Emergency Preparedness, 1972. *Disaster Preparedness:* 1, 3: Washington, D.C.

18. Youd, T.L., D.R. Nichols, E.J. Helley, and K.R. Lajoie, 1975. Liquefaction potential. In R.D. Borcherdt (ed.), Studies for Seismic Zonation of the San Francisco Bay Region. *U.S. Geological Survey Professional Paper* 944: 68–74.

19. Office of Emergency Preparedness, 1972. *Disaster Preparedness:* 1, 2: Washington, D.C.

20. Slemmons, D.B., and C.M. DePolo, 1986. Evaluation of active faulting and associated hazards. In *Active Tectonics*. National Academy Press: Washington, D.C.

21. Hansen, W.R., 1965. The Alaskan earthquake, March 27, 1964: Effects on communities. *U.S. Geological Survey Professional Paper* 542-A.

22. Oppenheimer, D., G. Beroza, G. Carver, L. Dengler, J. Eaton, L. Gee, F. Gonzales, A. Jayko, W.H. Li, M. Lisowski, M. Magee, G. Marshall, M. Murray, R. McPherson, B. Romanowicz, K. Sataker, R. Simpson, P. Somerville, R. Stein, and D. Valentine, 1993. The Cape Mendocino, California, earthquakes of April 1992: subduction at the triple junction. *Science,* 262: 433–438.

23. Radbruch, D.H., and B.J. Lennet, 1966. Tectonic creep in the Hayward fault zone, California. *U.S. Geological Survey Circular* 525.

24. Steinbrugge, K.V., and E.G. Zacher, 1960. Creep on the San Andreas fault. In R.W. Tank (ed.), *Focus on Environmental Geology*. Oxford University Press: New York.

25. Pakiser, L.C., J.P. Eaton, J.H. Healy, and C.B. Raleigh, 1969. Earthquake prediction and control. *Science,* 166: 1467–1474.

26. Evans, D.M., 1966. Man-made earthquakes in Denver. *Geotimes,* 10: 11–18.

27. Schwartz, D.P., and K.J. Coppersmith, 1984. Fault behavior and characteristic earthquakes: Examples from the Wasatch and San Andreas fault zones. *Journal of Geophysical Research,* 89: 5681–5698.

28. Sieh, K., 1981. A review of geological evidence for recurrence times of large earthquakes. In R.H. Sibson and P.G. Richards (eds.), *Earthquake Prediction, An International Review.* M. Ewing Series 4, American Geophysical Union: Washington, D.C.

29. Scholz, C.H., 1990. *The mechanics of earthquakes and faulting.* Cambridge University Press: Washington, D.C.

30. Swan, F.H., 1988. Temporal clustering of paleoseismic events on the Oued Fodda fault, Algeria. *Geology,* 16: 1092–1095.

31. Press, F., 1975. Earthquake prediction. *Scientific American,* 232 (1): 14–23.

32. Sibson, R.H., 1981. Fluid flow accompanying faulting: field evidence and models in earthquake prediction. In R.H. Sibson and P.G. Richards (eds.), *Earthquake Prediction, An International Review.* M. Ewing Series 4, American Geophysical Union: Washington, D.C.

33. Slemmons, D.B., 1982. Determination of design earthquake magnitudes for microzonation. In *3rd International Earthquake Microzonation Conference Proceedings.*

34. Wells, D.L., and K.J. Coppersmith, 1994. New empirical relationships among magnitude, rupture length, rupture width, rupture

area and surface displacement. *Bulletin of the Seismological Society of America,* 84(4): 974–1002.

35. Auersch, L., 1988. Seismic response of three-dimensional structures using a Green's function approach. In A. Vogel and K. Brandes (eds.), *Earthquake Prognostics.* Friedr. Vieweg & Sohn: Brannschweig/Weisbaden: 393–403.

36. Damrath, R., 1988. Model response analysis of structures. In A. Vogel and K. Brandes (eds.), *Earthquake Prognostics.* Friedr. Vieweg & Sohn: Brannschweig/Weisbaden: 405–426.

2

Landforms, Tectonic Geomorphology, and Quaternary Chronology

TECTONIC GEOMORPHOLOGY

Landforms are surficial features, the collection of which constitutes the **land-scape.** Examples of landforms include large features, such as mountains and plateaus, as well as smaller features, such as alluvial fans, hills, canyons, slopes, and sand dunes (see discussion of Landscape Scale in Chapter 9, especially Table 9.1). **Geomorphology** is the study of the nature, origin, and evolution of the landscape, focusing on physical, chemical, and biological processes that produce or modify landforms.

It is important to understand the relationships between geologic environment and the landscape. Geological factors are important because landform development is closely related to the underlying structure of the Earth. In geo-morphology, structure is broadly defined to include rock and soil types, nature and abundance of fractures in the rocks, and faults and folds. Landform development also depends on the nature of surficial and geologic processes, including weathering (physical and chemical), fluvial (most of the landscape is produced by subaerial erosional and depositional processes related to streams and rivers), glacial, eolian (wind), mass wasting [including slope failures (landslide, mud-flow, and earthflow)], tectonic (plate motion, faulting, folding, tilting, uplift, and subsidence), and volcanic.

Understanding geomorphology, and thus landscape evolution, is facilitated by the use of **process-response models,** which are qualitative and quantitative representations of how processes influence landform development. For example, models have been developed to explain changes in deposition on alluvial fans that result from tectonic and fluvial processes and changes in climatic conditions [1]. Such models are tools for understanding tectonic processes because present-day alluvial fan morphology can be used to infer tectonic activity in the past and predict activity in the future. The process-response model for alluvial fans is dis-cussed in detail in Chapter 9.

Tectonic geomorphology may be defined in two ways: (1) the study of landforms produced by tectonic processes, or (2) application of geomorphic prin-ciples to the solution of tectonic problems. The first definition implies that we are interested in the landforms themselves—their shape and origins—as func-tions of tectonic processes. The second definition has a utilitarian value; it allows us to use geomorphology as a tool to evaluate the history, magnitude, and rate of tectonic processes.

Geomorphology is a valuable tool in tectonic investigations because the **geomorphic record,** defined as the set of landforms and Quaternary deposits present at a site or area, generally encompasses the last few thousand to about two million years. Investigation of the geomorphic record provides the basic data necessary to understand the role of active tectonics in the development of a site or an area. For example, study of stream channels offset by faulting and faulted Quaternary deposits may reveal the amount of displacement and timing of the last few earthquakes at a particular site—information that is critical in evaluating future earthquake hazards.

GEOMORPHIC CONCEPTS

The study of tectonic geomorphology requires investigation of geomorphic processes, tectonic processes, and earth materials, as well as an appreciation of how the landscape is formed and maintained and how it evolves through time.

Four important geomorphic concepts directly applicable to the study of tectonic geomorphology are:

- Landscapes evolve through time, and changes in the assemblage of landforms are predictable.
- During landscape evolution, abrupt changes may occur as thresholds are exceeded [2].
- The interaction of landscape evolution with thresholds results in complex processes; this interaction is called **complex response** [2].
- A change in form implies a change in process.

Landscape Evolution

William Morris Davis put forward a simple model of landscape evolution in the late 1890s called the **"cycle of erosion."** His model was based on the assumption that brief pulses of uplift are followed by much longer periods of tectonic inactivity and erosion [3]. Davis envisioned that during this cycle, the landscape would go from a brief period of **youth,** characterized by deep V-shaped valleys and an abundance of waterfalls and rapids, to a longer period of **maturity,** characterized by a great variety of landforms, including meandering rivers and floodplains, followed by a long period of **old age,** in which the landscape might be worn down to an erosional surface of low relief that he called a **peneplain** (Figure 2.1). The time necessary for this hypothetical cycle is several million years, and the cycle could be interrupted at any time by renewed uplift, called **rejuvenation.**

Over shorter periods of time (on the order of a few thousand years), landscapes may approach a **dynamic equilibrium** in which the form of the landscape is nearly independent of time [4]. However, it has been argued that variables such as climate, tectonic processes, and geomorphic processes may change too rapidly for equilibrium to be reached for any appreciable length of time [5].

There is no doubt that landscape evolution is occurring over a range of time spans. Change is the norm, and sometimes that change may be very rapid. For example, less than 100 k.y. is necessary to produce a coastline characterized by several broad, uplifted, marine platforms (such as along the coasts of New Zealand, Japan, and California; see Chapter 6). Over a similar span of time, composite fault scarps up to 100 m high also may develop. Finally, fault scarps produced by individual earthquakes are created and eroded away over periods of only a few thousand years. Thus, our discussion of geomorphic evolution comes to the conclusion that the landscape is constantly changing over variable scales of time and space. What is important is that the changes often are predictable, and this provides a mechanism for evaluating active tectonics.

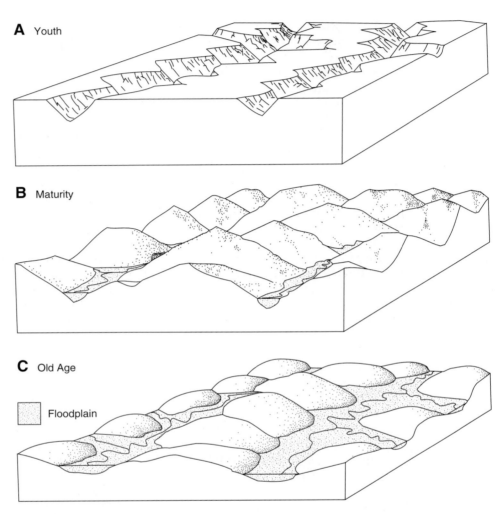

A Youth

B Maturity

C Old Age

Floodplain

FIGURE 2.1
The cycle of erosion as envisioned by William Morris Davis.

Thresholds

The life of a landscape must be like that of an airline pilot—long periods of routine work punctuated by brief periods of intense activity and change, like takeoff and landing or flying through an intense storm. Some landform changes reflect changes in **extrinsic variables**—external processes such as sudden earthquakes or the arrival of humans. Other changes result from changes in **intrinsic variables,** processes that are part of the everyday operation of the landscape system, as, for example, weathering, or movement of groundwater through deposits on a slope. Although a system may be held steady by a rough dynamic equilibrium, small changes may occur (for example, subsurface weathering that reduces the strength of rock and soil) until a **threshold** is surpassed and the balance collapses. When a threshold is exceeded, the effects may be both dramatic and quick. For example, in the Earthquake Cycle, the slow and steady deformation of rocks accumulates strain at a fault until the strain exceeds the strength of the rocks and an earthquake occurs.

Thresholds resulting from tectonic processes may change the operation of the system from deposition to erosion. For example, if a fault block containing a mountain and alluvial fans is slowly rotated or tilted, then the alluvial fans become steeper. Eventually a threshold slope is crossed, and streams on the alluvial fans downcut rapidly, moving the deposition of alluvium farther out from the mountain front. Thus, the processes near the mountain front change from deposition to erosion, and the deposition of alluvial-fan deposits is moved downstream [1].

An important consequence of understanding thresholds is that we are no longer limited to models in which long periods of slow change are interrupted by rapid external events (perturbations) such as climatic change and tectonic uplift. Rather, some changes in process and/or form, although they are abrupt, result from slower change that has occurred over a period of time within the system itself [2].

Complex Response

We have established that landscape evolution occurs and that changes in landforms are expected and predictable. We also have suggested that some of these changes occur suddenly as thresholds are crossed. The interaction between these two can lead to a complex evolution of a particular landscape. This complexity is often called **complex response** [2]. For example, we once thought that if several river terraces were present in a river valley, then each terrace must reflect some external event such as a climatic change or an uplift event. However, our understanding of geomorphic thresholds and complex response suggests that one perturbation or uplift event may cause a complex response in which several river terraces are produced. As a result, two drainages side by side may have different forms and histories. One basin may have several river terraces, while the adjacent basin has only one.

Changes in climate also have profound influences on geomorphic systems. For example, greater precipitation or aridity may result in channel deposition or downcutting, producing terraces. Climate changes, however, should be reflected in drainage basins on at least a regional scale. The fact that this is not always the case suggests that something else is occurring. That something else is a complex response. Figure 2.2 shows a drainage basin, the lower part of which has a floodplain. Faulting at the mountain front initiates a complex sequence of incision followed by deposition. Figure 2.2A shows the original

FIGURE 2.2
Idealized diagram showing the
concept of the complex
response. One tectonic event
of uplift produces two terraces.
See text for explanation.

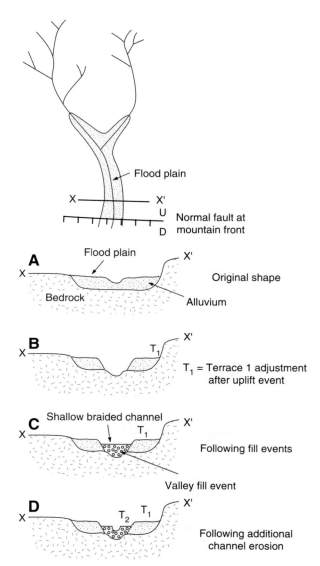

shape of the valley and floodplain prior to any disturbance by faulting. Following faulting, which uplifts the basin by several meters, the stream responds by
downcutting (erosion), producing terrace T_1 shown on Figure 2.2B. The downcutting and erosion produce sediment. As the erosion works its way headward,
the channel in the lower part of the basin adjusts to the change prior to the
upper part of the basin, where erosion is still occurring. By the time headward
erosion in the channel reaches the upper part of the basin, additional sediment
is transported downstream and deposited in the valley, perhaps causing a shal-

low, braided channel to form, as shown on Figure 2.2C. As deposition continues, the slope of the valley and braided channel increases until a threshold is crossed and downcutting occurs again, producing terrace T_2 (Figure 2.2D). Thus, in this simple example, two river terraces are produced by one tectonic disturbance.

Because no two river basins are exactly alike, the timing of erosion and deposition varies; thus the sequence of terraces may be variable. A climatic perturbation that resulted in downcutting would produce a similar sequence of events. This is the essence of complex response [2]. Thus, we must be careful when interpreting landforms in tectonically active areas. In places, multiple terraces are produced by multiple tectonic events; elsewhere, multiple terraces might be related by complex response to a single tectonic or climatic perturbation.

At the southern end of the San Joaquin Valley in California, San Emigdio Canyon crosses an axis of active uplift caused by a buried reverse fault (Figure 2.3). Upstream and downstream of the axis of uplift, there are two prominent Holocene stream terraces. Where the stream crosses the axis of uplift at the mountain front, however, there are several additional terraces, each separated by approximately a meter. The terraces upstream and downstream of the uplift probably reflect the more regional changes that have occurred. The additional terraces where uplift is centered may be related directly to earthquakes that have occurred at the mountain front. A feature common to many geomorphic systems is that in areas with the highest rate of uplift, not only are the *elevations* of landforms greater, but the *number* of episodic events recorded is greater.

Relationship Between Form and Process

A fundamental principle of geomorphology is that a change in landscape form often implies a change in landscape process. If you are walking up an alluvial fan with a uniform slope and you suddenly come to a sharp increase in gradient (perhaps to a near-vertical slope), followed by a return to the original slope, then this implies a change in process. In geomorphic terminology, the steep part of the slope is a **scarp,** possibly produced by erosion (called an erosional scarp) or tectonic processes such as faulting (in which case it is a **fault scarp**). The scarp also might be depositional in origin, representing the steep nose of a debris flow on the fan. Whatever the case, the change in form of the topography implies a change in the processes which formed it. Using this principle, geomorphologists observe landforms and look for anomalies that might reflect changes in process. When we are interested in tectonic geomorphology, we are often looking for landform surfaces that have been warped, tilted, uplifted, fractured, or otherwise deformed.

The principle that relates change in form to change in process is applicable at a variety of scales, from small fault scarps to mountain ranges. Just as the morphology of a fault scarp reflects a difference between erosion at its crest, producing

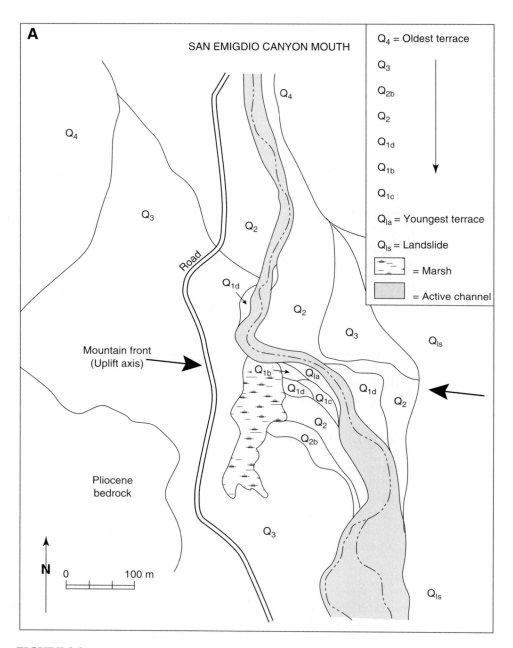

FIGURE 2.3
(A) Terraces near the mouth of San Emigdio Canyon, California. Multiple terraces have formed near the uplift axis associated with the buried Wheeler Ridge fault.

FIGURE 2.3 *(continued)*
(B) Photograph of terraces at San Emigdio Canyon.
(After Laduzinsky, 1989. M.A. thesis, University of California, Santa Barbara)

a convex slope, and deposition at its base, producing a concave slope, so too is the morphology of an eroding mountain block fundamentally different from the morphology of the adjacent basin into which material eroded from the mountain range is deposited.

TECTONIC GEOMORPHOLOGY AND FAULTING

Active faulting causes a variety of landform features, including fault scarps, warped and tilted ground, subsidence features such as sag ponds, and offset features such as stream channels. Each major category of faulting—strike-slip, normal, and reverse—may be discussed in terms of a characteristic assemblage of landforms. There is a fair amount of overlap between these different assemblages because many faults have oblique displacement, partially strike-slip and partially vertical. The balance between strike-slip and vertical displacement may vary significantly on a given fault system. Nevertheless, because specific processes tend to produce a particular set of responses, and therefore a particular assemblage of landforms, a generic classification of landforms is possible.

Landforms of Strike-Slip Faulting

The characteristic assemblage of landforms produced by active strike-slip faulting is shown on Figure 2.4. These features include:

- **Linear valleys** are troughs along main fault traces. These often develop because continued movement along recent fault traces crushes the rock, making it more vulnerable to erosion. Streams commonly follow these zones of weakness and flow some distance along the troughs.
- **Deflected streams** are streams that enter a fault zone at an oblique angle and flow parallel to the fault for some distance before returning to the original orientation of flow. Streams may be deflected either in a right or a left sense.
- **Offset streams** are streams displaced by faulting; they indicate the direction of relative displacement. The offset may reflect cumulative off-set of several earthquakes. Eventually the stream may erode a more

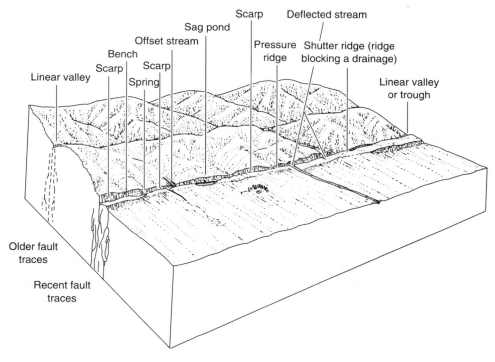

FIGURE 2.4

Assemblage of landforms associated with active strike-slip faulting.
(After Wesson et al., 1975. *U.S. Geological Survey Professional Paper* 941A)

direct route across the fault zone, producing a **beheaded stream** at the fault trace (Figure 5.8 shows an example of an offset stream).

- **Shutter ridges** (Figure 2.5) are formed where a fault displaces topography and moves ridge crests on one side of the fault against gullies on the other side. Because this topography is commonly carved by streams running perpendicular to faults, shutter ridges can be closely related to deflected and offset streams.
- **Scarps** can be produced by strike-slip motion by two possible mechanisms: (1) a small component of vertical displacement on individual fault strands results in local vertical separations, or (2) topographic relief on displaced landforms results in fault-parallel scarps.
- **Sag ponds** are often found in the fault zone and generally are related to downwarping between two strands of the fault zone (Figure 2.6).
- **Springs** are often found along the fault zone because the crushed rock associated with faulting can be either an effective barrier to, or conduit of, groundwater, forcing it to the surface. Along the southern San Andreas fault in the Coachella Valley, California, fault traces are often delineated by native palm trees that utilize the shallow groundwater. In some instances, the springs and palms form small oases along the fault (Figure 2.7).

FIGURE 2.5
Shutter ridges along the San Andreas Fault.
(Photograph by E.A. Keller)

FIGURE 2.6
Sag pond (dry) along the San Andreas fault in central California. Note that the line down the center of the depression is a fence, not a fault-related feature.
(Photograph by E.A. Keller)

- **Benches** consist of elevated, relatively flat topography in strike-slip fault zones. These benches may be slightly warped or tilted. These features usually are due to the displacements between several fault segments or strands in the zone.
- **Pressure ridges** are small warped areas produced by compression between multiple traces in a fault zone (Figure 2.8). Where a fault subsequently breaks through previous pressure ridges, shutter ridges may be formed.

Within strike-slip fault zones, simple shear may produce a variety of structures and forms, including fractures, folds, normal faults, thrust faults, and reverse faults (Figure 2.9). In Figure 2.9, the dashed circle is an imaginary area that is deformed by simple shear into the ellipse shown. As a result of that deformation, both contraction and extension occur. Extension produces normal faults and **grabens** (fault-bounded basins), sometimes at a microscale of only a few meters. Contraction produces reverse faults and small folds in the fault zone. Simple shear also produces **synthetic** and **antithetic shears** (faults with the same and opposite sense of displacement as the master fault, respectively) as well as what are called P-shears. Orientations of those fractures are also shown on Figure 2.9. The landforms and structures associated with simple shear are

FIGURE 2.7
Palm trees mark the trace of the San Andreas fault in Indio Hills, California.
(Photograph by E.A. Keller)

FIGURE 2.8
Pressure ridges along the San Andreas fault in central California.
(Photograph by E.A. Keller)

A

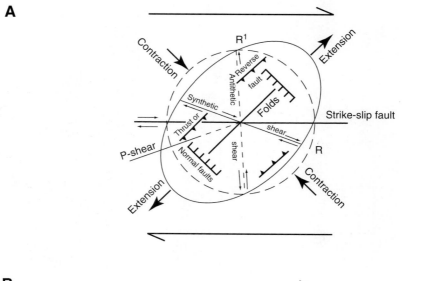

B

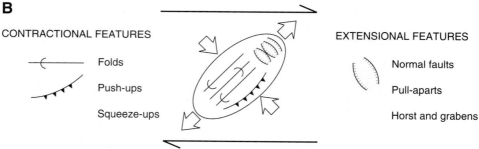

CONTRACTIONAL FEATURES

    Folds

    Push-ups

    Squeeze-ups

EXTENSIONAL FEATURES

    Normal faults

    Pull-aparts

    Horst and grabens

FIGURE 2.9
Simple shear associated with active strike-slip faulting produces a variety of fractures, faults, and folds with characteristic orientation (A), as well as extensional and contractional features (B).
(After Wilcox et al., 1973. *American Association of Petroleum Geologists Bulletin*, 57: 74–96; Sylvester and Smith, 1976. *American Association of Petroleum Geologists Bulletin*, 60: 2081–2102)

most common within fault zones—which, for large faults, may be a kilometer wide—but in some instances may extend beyond the fault zone.

    Indio Hills, east of Palm Springs, California, is an area along the San Andreas fault where many fault-related landforms are present. Figure 2.10 shows an offset alluvial fan where a number of fault-related landforms can be seen. These include shutter ridges, deflected streams, beheaded streams, sags, and microtopography [consisting of a small **horst** (fault-bounded uplifted block) and **graben** (fault-bounded depression; in this case, small downdropped blocks produced by normal faulting)] with orientations consistent with the simple shear model of Figure 2.9.

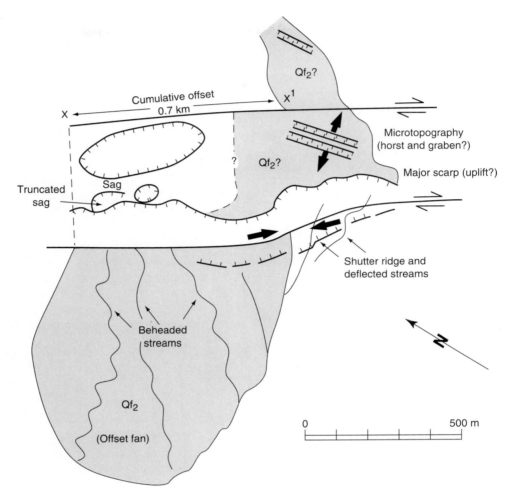

FIGURE 2.10
Offset alluvial fan and related landforms along the San Andreas fault, Indio Hills, California.
(After Keller et al., 1982. *Geological Society of America Bulletin,* 93: 46–56)

Figure 2.11 shows a low-sun–angle photograph of the offset alluvial fan and a 30-m high scarp associated with a small left bend of the main fault trace (see Figure 2.10). The left bend is thought to be responsible for the vertical component of motion at this point along the fault which produced the scarp. Localized uplift and subsidence are associated with bends and steps in a strike-slip fault (Figure 2.12). A left bend in a right-lateral fault system produces an area of uplift and thus is called a **restraining bend.** A right bend in a right-lateral fault produces an area of subsidence and thus is called a **releasing bend.** If two or more approximately parallel

FIGURE 2.11
Offset alluvial fan, Indio Hills, California. See Figure 2.10.
(Photograph courtesy of Woodward-Clyde consultants)

fault traces in a strike-slip fault step to the left or right, then these are called **restraining steps** and **releasing steps,** respectively, and also can produce areas of uplift or subsidence [6, 7].

Strike-slip faulting is not limited to the land. Figure 2.13 shows part of the southern California borderland not far offshore from San Diego, where the San Clemente fault cuts the ocean floor [8]. This large, right-lateral strike-slip fault parallels the San Andreas fault system and has many of the landforms we see on land, including linear troughs, sags, fault scarps, offset or deflected channels, and large tectonic benches. Some of the higher scarps have relief of about 200 m.

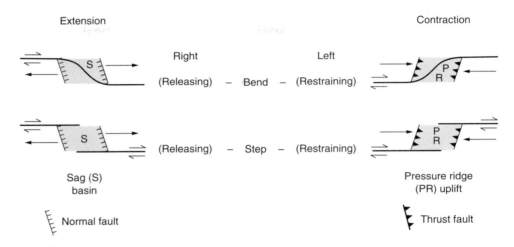

FIGURE 2.12
Sags and pressure ridges associated with bends and steps along strike-slip faults.
(After Crowell, 1974 [6]; Dibblee, 1977 [7])

Restraining and releasing bends also are found at more regional scales, producing much larger basins or areas of uplift. The general trend of the San Andreas fault is northwest-southeast, but near Los Angeles a section of the fault several hundred kilometers long is oriented closer to east-west (Figure 2.14). The western end of this anomalous east-west section of the fault is a sharp bend known as the "Big Bend." The geometry of the east-west section of the fault, sandwiched between the two northwest-southeast–trending sections, forms a large left bend similar to that shown on Figure 2.12, and as such should be associated with contraction. In fact, the Western Transverse Ranges lie along the Big Bend; they are east-west–trending mountains, in contrast to the north-west-southeast–trending Peninsular and Coast Ranges that characterize much of the California coast. Uplift of the Transverse Ranges can be explained in part as compression caused by this large left bend of the San Andreas fault. It is also possible that the San Francisco Bay is a pull-apart basin (produced by releasing bends or steps) related to several strands of the San Andreas fault in that region. Discussion of some of the large-scale features associated with strike-slip faulting is a speculative but interesting area for future work.

Landforms of Normal Faulting

Some of the most remarkable topography on land and beneath the oceans is associated with crustal extension and normal faulting. The mid-ocean ridges are the longest and most continuous mountain chains on Earth. The axes of oceanic ridges are marked by rift valleys, typically bounded by large normal faults. Rift

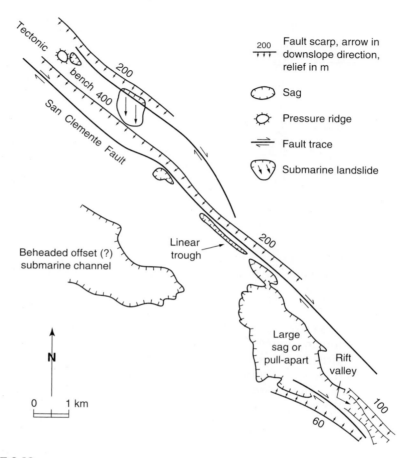

FIGURE 2.13
Idealized map of part of the San Clemente fault zone. Data are from Seabeam survey.
(Courtesy of M.R. Legg, University of California, San Diego)

valleys are also found on a smaller scale on continents (Figure 2.15). Perhaps the best example of continental rifting is the East African Rift system that extends for thousands of kilometers (Figure 2.16). The major landform of the system is the rift valley, which is a graben. The morphology of the East African Rift system is similar to that found along some oceanic ridge systems, where overlapping segments of ridges have a basin between them. The fault scarps produced by normal faulting in rift valleys, both on land and beneath the ocean, are some of the largest fault scarps found on Earth. The East African Rift system, like the oceanic ridge systems, is a topographically high area with active volcanism, including the spectacular Mt. Kilimanjaro and other peaks. The rift valleys of the East African rift are segmented along their lengths into a number of asymmetric

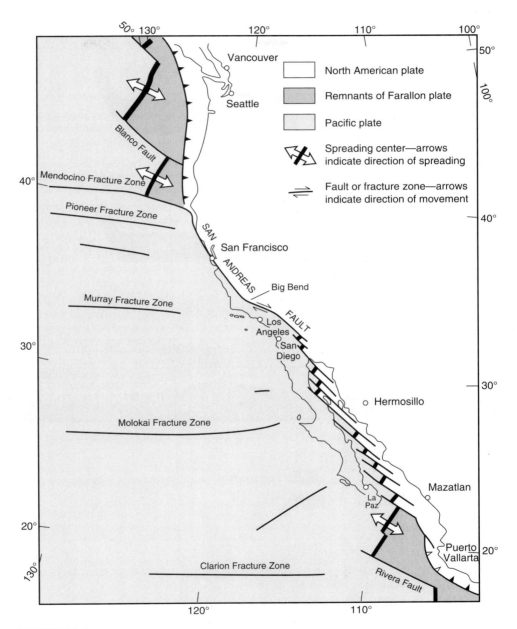

FIGURE 2.14
Tectonic framework of West Coast United States.
(After Irwin, 1990. *U.S. Geological Survey Professional Paper* 1515)

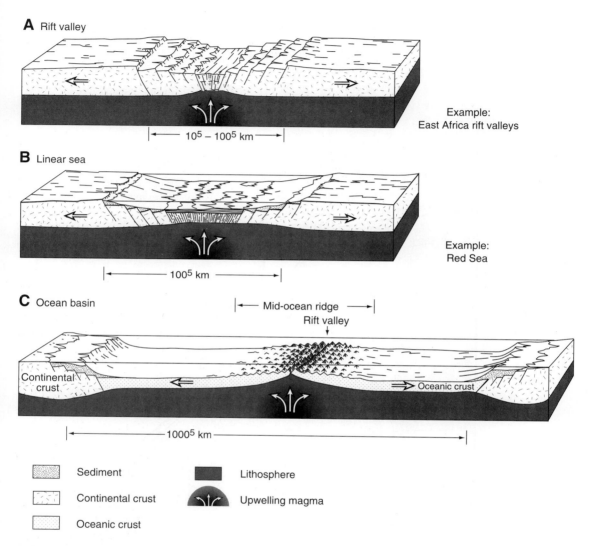

A Rift valley

Example:
East Africa rift valleys

$10^5 - 100^5$ km

B Linear sea

Example:
Red Sea

100^5 km

C Ocean basin

Mid-ocean ridge

Rift valley

Continental crust

Oceanic crust

1000^5 km

Sediment

Continental crust

Oceanic crust

Lithosphere

Upwelling magma

FIGURE 2.15
Landforms produced when the crust is pulled apart. Rifting and sea-floor spreading produce rift valleys, linear seas, and ocean basins.
(After Lutgens and Tarbuck, 1992. *Essentials of Geology.* Macmillan: New York)

basins (half grabens in which the valley is downdropped along a primary set of normal faults on one side of the valley) with lengths on the order of 100 km. Some of these basins are the sites of large lakes, such as Lake Tanganyika and Lake Victoria. The tectonic geomorphology of the East African rift valley is known in a gross sense, but much more basic research is necessary to understand the rates of tectonic processes and how they have produced this complex landscape. Understanding the geomorphic history of continental rift valleys may provide important information about how their larger cousins, the oceanic rifts, formed. The oceanic rifts represent the later stages of evolution of continental rift valleys, subsequent to the breakup of continents. Studies of the oceanic rifts are frustrated by the difficulty in obtaining topographic data and images comparable to those available on land.

In the United States, topographic expression of normal faulting is nowhere better expressed than in the Basin and Range province (Figure 2.17). The gross topography of the region has been compared to an army of caterpillars marching south from Idaho, or north from Mexico, depending on your bias. The caterpillars are long mountainous spines bounded by steep normal faults. The intervening basins (fault-bounded depressions) are grabens.

Two types of normal faults are shown in Figure 2.17B: high-angle normal faults that bound the basins and ranges (producing steep, linear mountain fronts), and a very-low-angle normal fault known as a detachment fault. Uplift of the ranges of the Basin and Range has been accompanied by block tilting, illustrated in Figure 2.17B. This tilting includes both down-to-the-east and down-to-the-west rotations, with regions of consistent tilt polarity separated by distinct "accommodation zones" [9]. Tectonic stress and crustal thinning responsible for the normal faulting in the Basin and Range began several million years ago and is going on today.

Normal faults near the surface are generally steeply dipping (~60°). As a result, mountain fronts along active normal faults tend to be straight and steep. However, there are exceptions to this; for example, mountain fronts at Dixie Valley and Pleasant Valley, Nevada, are more sinuous because the normal fault zones are themselves sinuous. The same is true for the mountain front at Salt Lake City, Utah [10]. The combination of vertical motion on range-bounding normal faults and stream incision in the valleys results in formation of triangular facets. Triangular facets are roughly planar surfaces with broad bases and upward-pointing apexes that occur between valleys that drain the mountains. Often there is a series of such features, sometimes called "flatirons," as shown on Figure 2.18. Triangular facets are characteristic of mountain fronts associated with active normal faults.

The western boundary of the Basin and Range is the Sierra Nevada of California. A system of normal and strike-slip faults lies along the eastern flank of the range. Just north of the town of Bishop, California, hundreds of fault scarps produced by normal faulting cut what is known as the Volcanic Tableland. The Tableland was produced about 750 ka by a catastrophic volcanic eruption. Volcanic ash covered an area of at least 1000 km^2 to an average depth of 150 m. In the past 750 k.y., numerous earthquakes have broken that surface (Figure 2.19) forming the fault scarps. The Volcanic Tableland is noteworthy because the durable surface records many faulting events so well, but also because faulting over such a broad area is not common in the Basin and Range. The pattern is attributed to flexure of the crust, accommodating westward tilting west of the Tableland and eastward tilting east of the Tableland [11].

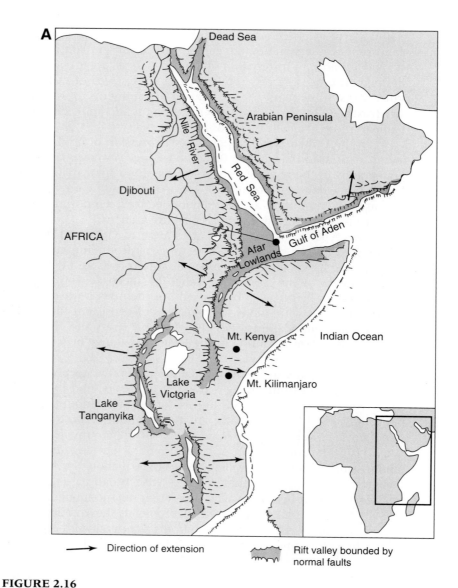

FIGURE 2.16

(A) East African Rift System.

(From Lutgens and Tarbuck, 1992. *Essentials of Geology.* Macmillan: New York)

Normal faults result from crustal extension and thinning, and this environment also favors volcanic activity. Thin crust often is associated with high heat flow, and partial melting of rocks at depths of a few kilometers produces magma to feed volcanic activity. Active volcanism is common along active spreading ridges

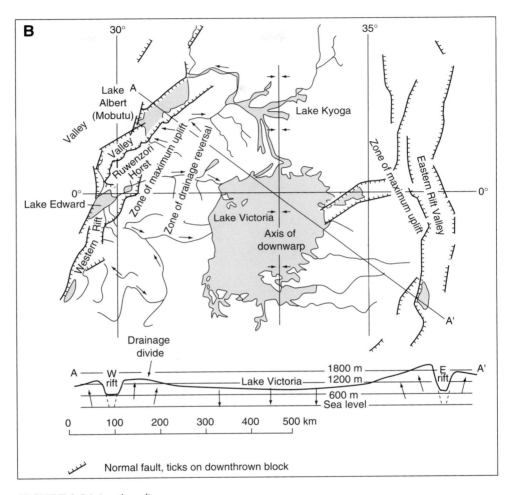

FIGURE 2.16 *(continued)*
(B) Tectonic framework and topographic section of part of the East African Rift System.
(After Bloom, A.L., 1991. *Geomorphology.* Prentice-Hall: Englewood Cliffs, NJ)

(mid-ocean ridges), rift valleys, and other locations associated with extensional tectonics (normal faulting), including the Basin and Range province. For example, in the Mono Basin, located north of the Volcanic Tableland, recent volcanism (within the last 40 k.y.) and intrusion of dikes reflect tensional strain previously taken up by normal faulting along the Sierra Nevada frontal fault system [12].

Normal faulting over several millions of years can form **escarpments** (long, nearly continuous cliff-like slopes) that may dominate the topography of a region. For example, normal faulting has dominated landscape development in much of eastern Greece and the Aegean Sea for at least the last several million years [13].

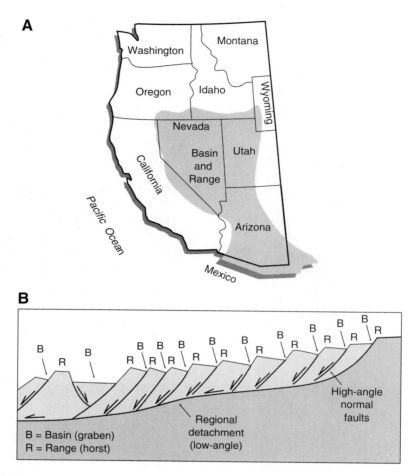

FIGURE 2.17
(A) Location of Basin and Range province, and (B) regional tectonic framework including high-angle normal faults that produce the basins and ranges, and regional low-angle detachment faults.

The Gulf of Corinth (Figure 2.20) is a graben bounded by normal fault escarpments with relief (difference in elevation from base to top of the escarpment) of several hundred meters. Two prominent escarpments south of Skinos climb 1100 m in two steps [13]. In February and March 1981, three moderate earthquakes (M = 6.7, 6.4, and 6.4) produced considerable ground rupture on normal faults along both the north and south margins of the Gulf of Corinth (Figure 2.20). The major surface ruptures occurred along a 10-km-long normal fault segment trending east-west (east of Pision, Figure 2.20) and following the base of a very prominent escarpment produced by numerous earlier faulting events. Vertical displacements

FIGURE 2.17 *(continued)*
(C) Photograph of Central Arrow Canyon Range in the Basin and Range province.
(Photograph © J. Shelton)

were as much as 1.5 m, but mostly ranged from 0.5 m to 0.7 m. Study of the ground rupture showed that this faulting was clearly the most recent displacement along the faults responsible for the major topographic features (escarpments) on the margin of the Gulf of Corinth [13].

In summary, the assemblage of landforms associated with normal faults includes steep, linear mountain fronts; fault scarps; horsts and grabens; escarpments; volcanic landforms (lava flows, cones, etc.); and at regional scales, rift valleys and axial rifts of oceanic ridge systems.

Landforms of Reverse Faulting

Reverse faulting generally is found in areas of crustal thickening, where mountains are being constructed. Some of the most spectacular scenery in the world is produced by uplift associated with reverse faulting. For example, at convergent plate boundaries, thrust faulting associated with subduction produces a

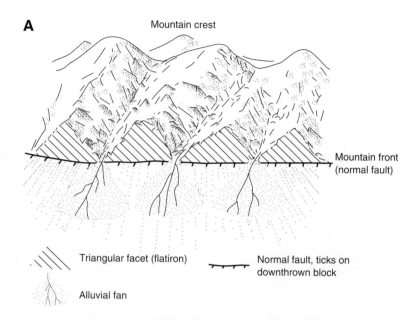

FIGURE 2.18
(A) Idealized oblique diagram showing a mountain front, normal fault, and triangular facets (flatirons) in the southern San Joaquin Valley, California. (B) Photograph of features shown in (A) looking across Saline Valley at Inyo Mountains and snowy crest of Sierra Nevada beyond. (Photograph © J. Shelton)

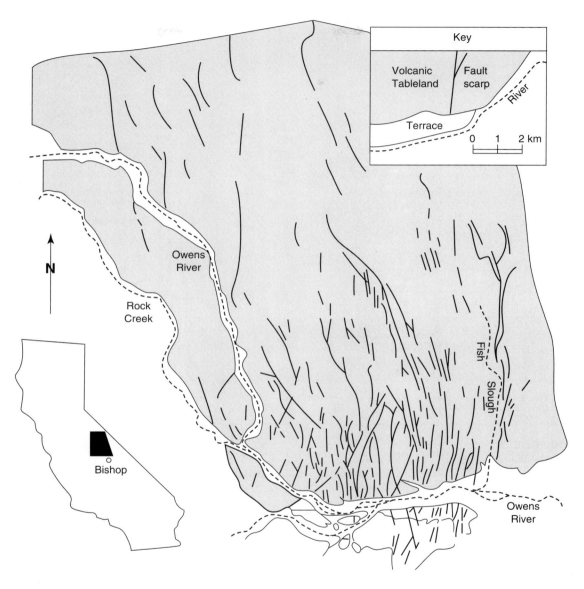

FIGURE 2.19
Fault scarps on the Volcanic Tableland, eastern California. The Tableland is a 738 ka volcanic sheet that has preserved extensive rupture across its surface.

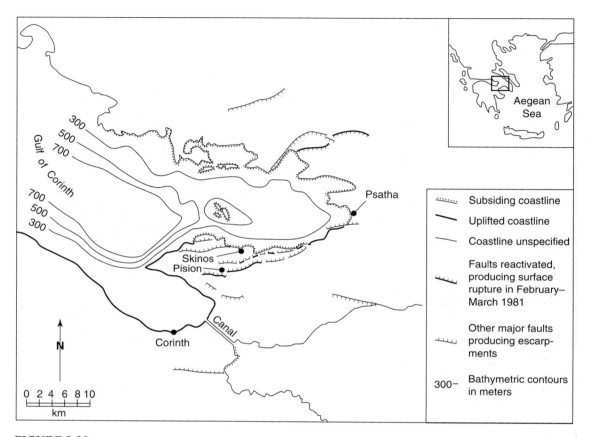

FIGURE 2.20
Normal faults in the Gulf of Corinth, Greece. Also shown are areas of surface rupture
from the 1981 earthquakes, subsiding coastlines, and uplifted coastlines.
(After Jackson et al., 1982. *Earth and Planetary Science Letters,* 57: 377–397)

variety of landforms, including uplifted marine terraces, anticlinal hills
(upwarps), and synclinal lowlands (downwarps). One of the best examples of
this is the landscape near Eureka, in northwestern coastal California, where
folds associated with the Cascadia Subduction Zone are exposed onshore. Landforms such as Humboldt Bay and other coastal lowlands are at sites of synclines, whereas intervening hills are anticlines (Figure 2.21) [14, 15]. The
important subject of active folding and fold-related landforms is discussed in
detail in Chapter 7.

     The assemblage of landforms associated with reverse faulting includes belts
of active folding and faulting called fold-and-thrust belts. Thrust faults are a low-
angle variety of reverse faults and often are associated with folds. Active folding
has produced spectacular folded topography in the Zagros fold belt of Iran (Figure

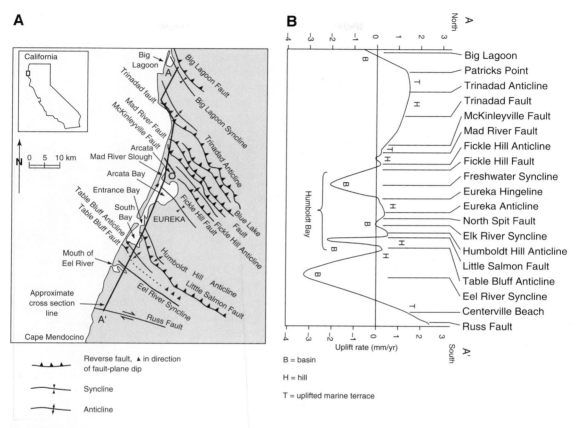

FIGURE 2.21
(A) Map of folds and thrust faults near Eureka, California. (B) Profile of uplift rates A–A'.
Uplift rates clearly match the topography. Anticlines and thrust faults produce hills and
uplifted marine terraces and synclines produce bays (basins).
(After Valentine, 1992 [15]; Clarke and Carver, 1992 [14])

2.22), and the northern Apennine Mountains in Italy, as well as in other parts of
the world. Landforms associated with reverse faulting include steep mountain
fronts, fault scarps, extensional features, and landslides. The 1980, $M = 7.3$, El
Asnam earthquake in Algeria produced a mean vertical displacement of 6 m on
a 30-km-long reverse-fault system. One of the surprising discoveries following
the earthquake was the amount of tensional (normal) faulting that accompanied
the event. Compressional deformation caused about 5 m of uplift of the anticlinal fold, but the most extensive surface-deformation features caused by the
earthquake were tensional, including normal fault scarps and grabens [16].
These normal faults are all on the upper plate above the fault, which was bent
during faulting, producing the extension (Figure 2.23).

FIGURE 2.22
Oblique photograph of an anticline in the active Zagros fold belt, Iran.
(Photograph courtesy of Aerofilms)

PLEISTOCENE AND HOLOCENE CHRONOLOGY

Geologists have subdivided the eons of geologic time in order to understand events in Earth history in their correct chronologic sequence and to correlate local sequences of rocks within their regional and global context (Table 2.1). For scientists and citizens concerned with active tectonics, the period of most interest is the **Quaternary,** which is the most recent 1.65 m.y., consisting of the **Pleistocene** and **Holocene Epochs.** The Pleistocene is divided into three parts.

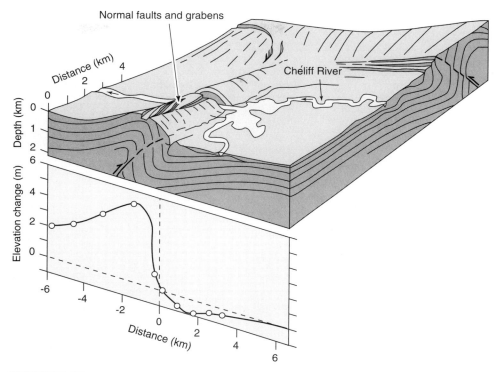

FIGURE 2.23
Idealized diagram of a fold deformed during the 1980 El Asnam earthquake ($M = 7.3$) and graph of surface uplift produced by the event. The fold was produced by a sequence of such events. Note the extensional features at the crest of the fold.
(After Stein and Yeats, 1989. *Scientific American,* 260(6): 48–57)

- Early Pleistocene, from approximately 1.65 Ma to 780 ka. The boundary at 780 ka coincides with the time of the most recent reversal of the Earth's magnetic field.
- Middle Pleistocene, from 780 ka to 125 ka. The boundary at 125 ka marks the most recent major interglacial and associated highstand of sea level.
- Late Pleistocene, from 125 ka to 10 ka. The boundary at 10 ka marks the end of the Pleistocene, and the beginning of the Holocene, and reflects a major climatic change. Often in active tectonics we are interested in the part of the Late Pleistocene from approximately 18 ka to 10 ka. The boundary at 18 ka marks the maximum of the most recent glaciation.
- Holocene, from 10 ka to present. The Holocene is the time period in which human civilizations burgeoned and grew. Across most of the planet, the climate stabilized (with some notable exceptions) to the conditions we know today. The early to mid-Holocene marks the Neolithic

TABLE 2.1
The geologic time scale.

Era	Period	Epoch	Millions of Years	
			Duration	**Before Present**
CENOZOIC	Quaternary	Holocene	0.01	
		Pleistocene	1.65	
MESOZOIC	Tertiary	Pliocene Miocene Oligocene Eocene Paleocene	64	
PALEOZOIC				66
	Cretaceous		78	
	Jurassic		64	
	Triassic		37	
				245
	Permian		41	
PRECAMBRIAN	Pennsylvanian		34	
	Mississippian		40	
	Devonian		48	
	Silurian		30	
	Ordovician		67	
	Cambrian		65	
				570
	Proterozoic		2500	
	Archean		3800?	

(From Palmer, 1983. *Geology,* 11: 503–504)

Revolution, which included such human advances as agriculture, settled communities, and development of the earliest complex societies. Active faults are defined as faults that have moved during the Holocene, and the study of the earthquake history during the Holocene allows us to estimate future earthquake hazards.

Dating is perhaps the most critical tool in assessing active tectonics, and it is often the most difficult part of a tectonic study. If there is no chronology (in particular, no numerical dates), then we cannot calculate rates of tectonic processes ("no dates, no rates"). It is usually much easier to measure deformation resulting from

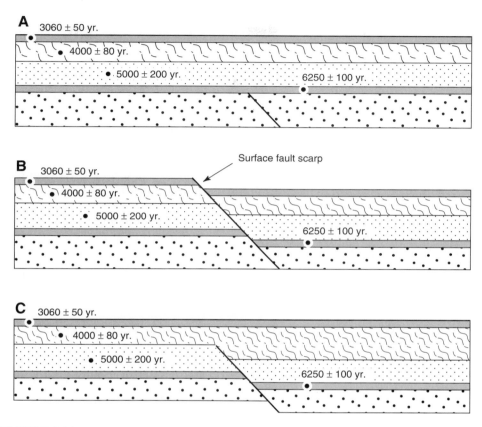

FIGURE 2.24
Principles of dating the most recent fault rupture. In all three cases, several horizons have been dated by [14]C. In case (A) the minimum age for the most recent faulting event is 6250 ± 100 yr. In case (B) the maximum age for the most recent faulting is 3060 ± 50 yr. In case (C) the most recent faulting occurred between 5000 ± 200 and 4000 ± 80 yr.

tectonic activity than it is to accurately date when that deformation occurred. For example, we can easily measure the amount of vertical displacement associated with a past earthquake, but unless we are able to date the deposits associated with the displacement, we cannot calculate the rate of activity. If we are studying a sequence of alluvial deposits and they overlie and are not displaced by a fault below, than dating those deposits provides a minimum age since the last faulting (earthquake) event (Figure 2.24A). If the deposits are displaced, then dating them provides a maximum age for the most recent faulting event [17] (Figure 2.24B). Under ideal conditions, we would like to have several dates at a particular location—one or more for the overlying deposits that are not faulted and others for deposits that are faulted (Figure 2.24C). Given this information, we can bracket the time in which displacement occurred. New techniques are now entering the geologist's toolbox for establishing the age of fault motion, including numerical dating of fault gouge and age of exposure of geomorphic surfaces.

Numerical age control for faulting events is vital in tectonic and earth-quake studies in order to establish three parameters: the age of the most recent rupture; the typical duration of time between events (recurrence interval); and the long-term displacement rate on the fault. The age of most recent ruptures, in particular, has both scientific and legal significance. In California, in order to call a fault "active" it must be demonstrated that the fault has moved in the past 10 k.y. [18]. For the U.S. Nuclear Regulatory Commission, a capable fault is one that has moved at least once in the past 50 k.y. (conveniently within the limits of conventional radiocarbon dating methods) or several times during the past 500 k.y. (see Chapter 1) [19].

At present, over 20 different methods of dating are useful for studying active tectonics. These methods may be categorized in terms of whether they are numer-ical methods (providing an absolute age in years), relative dating methods, or cor-relation methods [17]. Appendix A lists the more common methods used to date Pleistocene and Holocene materials and summarizes the age range over which each method can be applied, the basis for the method, and selected comments.

Historical records provide the most accurate chronology, but these are lim-ited to only about 200 years in the United States. In other areas such as China, the historical record extends several thousand years. Given the brevity of human history compared with geologic history, we are usually forced to resort to analyti-cal dating methods. All of the dating methods listed in Appendix A are applicable to active-tectonics studies under specific circumstances. Which method is used largely depends on the environment being studied. For example, if volcanic ash is present, then the logical methods are tephrachronology or potassium-argon analysis. On the other hand, if carbonate rinds are found on clasts in soil hori-zons, they may be dated by using the uranium-series method. If the rinds are of Holocene or latest Pleistocene age, then the ^{14}C method may provide some age control. [14]C is the most often utilized of the dating techniques, and also the most available, least expensive, and probably the most thoroughly tested. Standard radiocarbon analyses can date material as old as about 50 ka (but many scientists trust the results only as far back as about 35 ka). Accelerator mass spectroscopy (AMS) is a variety of ^{14}C analysis that increases the temporal range that is datable, shrinks the error bars, and decreases the minimum sample size. Additional com-plications have been revealed by the extensive use of radiocarbon analyses. A fundamental assumption of the ^{14}C method is that the reservoir of ^{14}C in the atmosphere, to which all terrestrial organisms are equilibrated, has remained unchanged through time. Independent dating techniques reveal that this assump-tion is slightly in error, and radiocarbon ages are systematically too young by up to 3.5 k.y. at 20 ka, for example [20]. Independent corrections exist for this sys-tematic error from tree rings [21], glacial varves (annual lake strata) [21], and uranium-thorium dates from corals [22].

In addition to systematic errors, random errors can be introduced into ^{14}C ages by sample contamination with carbon that is either too old or too young. Incorporation of 10% dead carbon (carbon with no ^{14}C left) into a sample will provide a date approximately 800 yr too old, and contamination of only about 0.5% recent carbon in any very old (>50 ka) sample will produce a date of about 40 ka, even if the other 99.5% of the sample formed millions of years ago [17].

In addition to methods of dating geologic deposits, a battery of new techniques has been introduced to directly date the duration that a landform has been exposed at the Earth's surface. Research is underway to evaluate the use of beryllium-10, aluminum-26, chlorine-36, and other isotopes, all of which are produced when cosmic rays interact with the Earth's atmosphere or its surface [17, 23]. Because these isotopes accumulate in measurable quantities in surface materials such as soils, alluvial deposits, and rock surfaces, their concentration is a potential measurement of minimum time of exposure of those surfaces. Therefore, in principle, it would be possible to measure the amount of beryllium-10 in the surface layer of a fault scarp in bedrock and determine the time the scarp has been exposed. This would be the duration of time since faulting.

Another new technique is being developed that may directly date the latest movement of a fault through analysis of the fault gouge. Electron spin resonance (ESR) takes advantage of the fact that ionizing radiation, from both cosmic rays and radioactive decay in surrounding material, occurs in the crystal structure of calcite or quartz [24]. Fault displacement can reset the ESR clock by either frictional heating or migration of fluids along the fault plane [25]. Thus any ESR signal measured in fault gouge should have accumulated only since the last fault rupture. That age can be estimated by measuring the signal and the in-situ radiation dose.

It is often difficult to collect useful dating material at a particular site where deformation from active tectonics occurs. As a result, it can be useful to establish a relative chronology based upon methods such as soil-profile development or rock and mineral weathering. The development of a soil chronology is a complex process that requires the description and analysis of many soil profiles [26]. The tool is, however, a powerful one because the chronology may be established over an entire region and then carried to an area where independent age control is not available. Soils on alluvial surfaces of a variety of ages may be studied to produce what is known as a **soil chronosequence,** which is a series of soils arranged from youngest to oldest in terms of relative soil-profile development. Some of the soils may be dated by radiometric and other methods at some location in the region and then, through correlation of the soils (comparing dated profiles with profiles of unknown age), the chronology may be applied to other sites where absolute dates are unavailable. For example, the offset alluvial fan along the San Andreas fault in the Indio Hills near Palm Springs (Figure 2.10) has a relatively weak soil profile development with an estimated age of 20 to 30 ka. This age was determined by comparing soil-profile development at the site to soils of known age in the region. Soil dates based on correlation with dated soils are only rough estimates, commonly with uncertainties of up to ± 25%. Applying the estimated

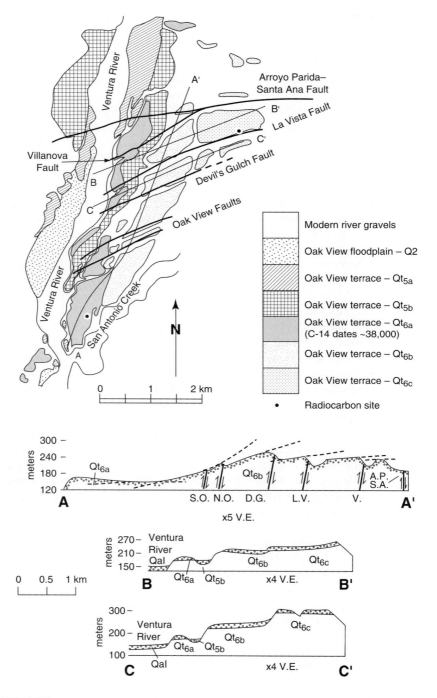

FIGURE 2.25
Faulted terraces of the Ventura River near Oak View, California. Cross section A–A' shows effects of faulting; cross sections B–B' and C–C' show the river terraces between faults. (After Rockwell et al., 1984 [27])

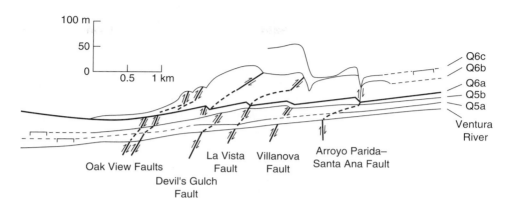

FIGURE 2.26
Longitudinal profiles of the modern Ventura River and faulted Late Pleistocene terraces.
Locations of faults shown on Figure 2.25.
(Modified after Rockwell et al., 1984 [27])

age range to the alluvial fan in the Indio Hills, which is offset about 700 m, provides a slip rate of approximately 23–35 mm/yr. As a second example of the use of soils in tectonic studies, consider the faulted terraces near Oak View, California, shown on Figures 2.25 and 2.26. Soils were described on the six river terraces. Development of a soil chronosequence allowed the faulted terraces to be correlated, and ^{14}C dates from charcoal in the terrace deposits for terraces Qt_{5a}, Qt_{5b}, and Qt_{6a} (along with measured displacement of the terraces) allowed calculation of slip rates for the faults (ranging from 0.3 mm/yr to 1.1 mm/yr). Ages for the older terraces (Qt_{6b} and Qt_{6c}, Figure 2.25) were estimated by extrapolation from the younger dated terraces [27].

In summary, development of the Pleistocene and Holocene chronology is probably the most difficult part of an active tectonics study. Nevertheless, it is necessary if the history and rates of activity are to be determined. Vigorous research underway is developing new methods of dating that will be more reliable in estimating ages of materials, surfaces, and events.

SUMMARY

Tectonic geomorphology is defined as the study of landforms produced by tectonic processes or the application of geomorphic principles to the solution of tectonic problems. Four important geomorphic concepts applicable to tectonic geomorphology are: (1) landscapes evolve through time, and changes in landforms are predictable, (2) abrupt changes in landscape evolution may occur as thresholds are exceeded, (3) interactions of landscape evolution with thresholds can produce complex response, and (4) a change in form implies a change in process. Active faulting produces a variety of landforms, and each major category of faulting may be discussed in terms of a characteristic assemblage of landforms.

Determination of the Pleistocene and Holocene chronology is critical to the solution of any tectonic-geomorphology problem. If there are no dates, there are no rates! Methods of establishing the chronology (dating) may be characterized in terms of whether they are numerical, relative, or based on correlation. Historical records provide the most accurate chronology but these are limited to only about 200 years in the United States. Vigorous research is now underway to develop new methods of dating.

REFERENCES CITED

1. Bull, W.B., 1964. Geomorphology of segmented alluvial fans in western Fresno County, California. *U.S. Geological Survey Professional Paper* 352-E.
2. Schumm, S.A., 1977. *The Fluvial System.* John Wiley & Sons: New York.
3. Davis, W.M., 1899. The geographical cycle. *The Geographical Journal,* 14 (5): 481–504.
4. Hack, J.T., 1960. Interpretation of erosional topography in humid temperate regions. *American Journal of Science,* 258A: 80–97.
5. Bull, W.B., 1975. Allometric change of landforms. *Geological Society of America Bulletin,* 86: 1489–1498.
6. Crowell, J.C., 1974. Origin of late Cenozoic basins in southern California. In W. Dickinson (ed.), Tectonics and Sedimentation. *Society of Economic Paleontologists and Mineralogists Special Publication* 22.
7. Dibblee, T.W., Jr., 1977. Strike-slip tectonics of the San Andreas fault and its role in Cenozoic basin evolvement. Late Mesozoic and Cenozoic Sedimentation and Tectonics in California, *San Joaquin Geological Society Short Course.*
8. Legg, M.R., and B.P. Luyendyk, 1982. Seabeam survey of an active strike-slip fault in the southern California continental borderland. *EOS: Transactions, American Geophysical Union,* 63: 1107.
9. Thenhaus, P.C., and T.P. Barnhard, 1989. Regional termination and segmentation of

Quaternary fault belts in the Great Basin, Nevada and Utah. *Bulletin of the Seismological Society of America,* 79: 1426–1438.
10. Bruhn, R., 1994, personal communication.
11. Pinter, N., and E.A. Keller, 1995. Geomorphic analysis of neotectonic deformation, northern Owens Valley, California: *Geologische Rundschau,* 84: 200–212.
12. Bursik, M., and K. Sieh, 1989. Rangefront faulting and volcanism in the Mono Basin, Eastern California. *Journal of Geophysical Research,* 93: 15,587–15,609.
13. Jackson, J.A., J. Gagnepain, G. Houseman, G.C.P. King, P. Papadimitriou, C. Soufleris, and J. Virieux, 1982. Seismicity, normal faulting and the geomorphological development of the Gulf of Corinth (Greece): the Corinth earthquakes of February and March, 1981. *Earth and Planetary Science Letters,* 57: 377–397.
14. Clarke, S.H., and G.A. Carver, 1992. Late Holocene tectonics and paleoseismicity, southern Cascadia subduction zone. *Science,* 255: 188–192.
15. Valentine, D.W., 1992. Late Holocene stratigraphy as evidence for late Holocene paleoseismicity of the southern Cascadia subduction zone, Humboldt Bay, California. Master's thesis, Humboldt State University, California.
16. King, C.G.P., and C. Vita-Finzi, 1981. Active folding in the Algerian earthquake of 10 October 1980. *Nature,* 292: 22–26.

17. Pierce, K.L., 1986. Dating methods. In *Active Tectonics*. National Academy Press: Washington, D.C.
18. State of California, 1973. California State Mining and Geology Board Classification (fault activity).
19. U.S. Nuclear Regulatory Commission, 1982. Appendix A. Seismic and geologic siting criteria for nuclear power plants. *Code of Federal Regulations – Energy,* Title 10, Chapter 1, Part 100. (App. A, 10, CFR 100).
20. Stuiver, M., and R. Kra, 1986. Calibration issues. *Radiocarbon,* 28: 805–1030.
21. Stuiver, M., 1970. Tree ring, varve, and carbon-14 chronologies. *Nature,* 228: 454–455.
22. Bard, E.B., R.G. Hamelin, and A. Zindler, 1990. Calibration of ^{14}C time scale over the past 30 ka using mass spectrometric U-Th ages from Barbados corals. *Nature,* 345: 405–410.
23. Pavich, M.J., 1987. Application of mass spectrometric measurement of ^{10}Be, ^{26}Al, ^{3}He to surficial geology. In A.J. Crone and E.M. Omdahl (eds.), Directions in Paleoseismology. *U.S. Geological Survey Open-File Report* 87-673.
24. Ikeya, M., 1975. Dating a stalactite by electron paramagnetic resonance. *Nature,* 255: 48–50.
25. Fukuchi, T., 1988. Applicability of ESR dating using multiple centres of fault movement—the case of Itoigawa-Shizuoka tectonic line, a major fault in Japan. *Quaternary Science Reviews,* 7: 509–514.
26. Birkland, P., 1984. *Soils and Geomorphology.* Oxford University Press: New York.
27. Rockwell, T.K., E.A. Keller, M.N. Clark, and D.L. Johnson, 1984. Chronology and rates of faulting of Ventura River terraces, California. *Geological Society of America Bulletin,* 95: 1466–1474.

3
Geodesy

INTRODUCTION

Geodesy is the science of the shape of the Earth, including both its general form and precise measurements of its surface. The first human attempts at accurate measurement of the land date to the first advanced civilizations. The Egyptians and Babylonians aspired to great works of architecture and engineering and required accurate surveys of the land to construct them. The sophistication and accuracy of surveying techniques have improved with the increasing ambitions of human civilization. In the modern world, networks of roads and rails, canals, power lines, and telecommunication lines unite and bind the planet, spanning even the most forbidding terrain. These works are built, figuratively speaking, on networks of highly accurate geodetic lines.

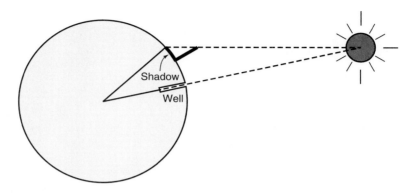

FIGURE 3.1
Eratosthenes observed that at noon on the summer solstice, the rays of the sun shone directly down a well in Aswan, but a pole in Alexandria cast a shadow. By measuring the length of the shadow and the height of the pole, and knowing the distance between the two sites, he was able to calculate the radius of the Earth.

Studies in active tectonics rely upon geodesy for measurements of almost imperceptible changes in the surface of the Earth, changes that signal ongoing tectonic activity. The first applications of geodesy were not so subtle. In Classical and Postclassical Greece, there was considerable debate whether the Earth was a flat disk or, as Pythagoras and Aristotle postulated, a sphere. The father of geodesy was Eratosthenes, who lived between 276 and 195 B.C. and served as the librarian of the Museum of Alexandria, Egypt. Eratosthenes observed that, at noon on the day of the summer solstice, the Sun shone directly down into a well at Aswan, along the Nile valley. At the same moment in Alexandria, farther north at the Nile delta, the Sun was not directly overhead, and a vertical pole cast a shadow (Figure 3.1). By measuring the length of the shadow and the length of the pole and knowing the distance between Aswan and Alexandria (5000 Egyptian stadia or 787.5 km), Eratosthenes calculated the radius of the Earth to be 6267 km [1, 2]. By this insightful observation and careful measurement, he not only proved that the Earth is roughly spherical in shape, but also that it is possible to estimate its size—an estimate that was within 2% of the true average radius of the Earth (6371 km)!

The spherical model of the Earth remained the governing paradigm for nearly two millennia after Eratosthenes, and it was the increasing accuracy of geodetic measurement that proved the model to be inadequate. In the early eighteenth century, it was found that a pendulum clock shipped from Paris to more equatorial latitudes ran slower near the equator. Sir Isaac Newton argued that this was evidence that the Earth was not truly spherical, but an **oblate ellipsoid,** bulging at the equator (Figure 3.2). The surface at the equator on an oblate body is farther from the center than is the surface at higher latitudes. Gravity would be fractionally weaker, and a pendulum would swing more

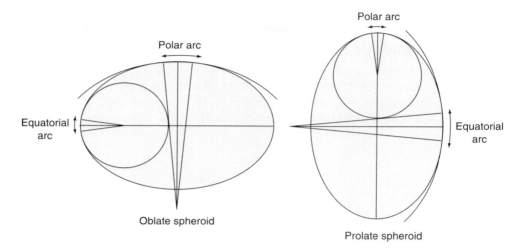

FIGURE 3.2
Eighteenth-century theories on the Earth's shape. English scientists believed that the planet was an oblate spheroid (left)—the shape of a pumpkin sitting on the porch at Thanksgiving. French scientists believed it was a prolate spheroid (right)—egg-shaped. On an oblate spheroid, the distance corresponding to one degree of latitude increases toward the poles; on a prolate spheroid, the distance diminishes from equator to poles. (After Fernie, 1991 [3])

slowly. Never avoiding an argument with the English, French scientists asserted that the Earth was **prolate**—bulging at the poles. The acid test of this dispute was to accurately measure the surface length of one degree of latitude, which would be greater near the poles on an oblate spheroid, but greater near the equator on a prolate spheroid. The French Academy of Sciences sent out two teams of geodesists, one to Ecuador and the other to Lapland. The scientists assigned to the southerly expedition surely had no idea what was in store for them—they were absent more than ten years and endured searing heat, icy cold, starvation, disease, and murder. The northern expedition fared somewhat better—despite heavy seas, attempted piracy, and a subzero Scandinavian winter, they returned within 18 months. Through very careful geodetic and astronomical observation, both expeditions succeeded in their objective. The French found the length of 1° of arc in Ecuador to be 340,404 ft, and in Lapland 344,532 ft, proving that the the English were right—the Earth is indeed oblate [3]. Subsequent remeasurement found that the northern expedition had overestimated the distance in Lapland by about 600 ft, but the conclusion was correct nonetheless. We now know that the radius of the Earth at the equator is 1 part in 298.25 (about 21.4 km) greater than at the poles. From a purely spherical point of view, the Mississippi River runs uphill—it is farther from the center of the Earth at its delta than at its headwaters.

PRINCIPLES OF GEODESY

Geodetic Frames of Reference

As the example of the Mississippi River illustrates, an important concept in geodesy is the frame of reference against which a position, including elevation, is measured. We now know that even the model of the Earth as an oblate spheroid is only an approximation (Figure 3.3). The global position of sea level, which is a

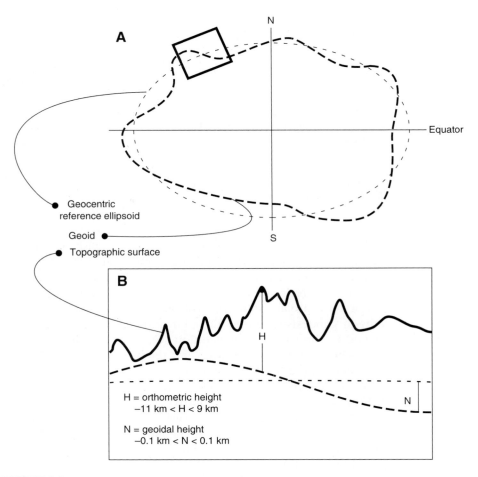

FIGURE 3.3

Relationship between the geocentric reference ellipsoid (cross section of an oblate spheroid), the geoid, and the surface topography of the Earth. Note that all deviations from a sphere are greatly exaggerated.

(After Wells, 1986. *Guide to GPS Positioning.* Canadian GPS Associates: Fredericton, NB)

surface of equal gravity potential known as the **geoid,** has an undulating shape that deviates from the oblate spheroid by as much as 100 m (Figure 3.3A). Elevation, as we commonly use the term, is a distance above or below the geoid, a distance more correctly called **orthometric height** (Figure 3.3B). Geodetic networks must be tied to the absolute vertical datum of the geoid, usually at several different locations. Mean sea level at a particular location is typically established by use of a **tide gauge** (see Chapter 6), which measures and averages the water level over long periods of time [4].

Geodetic Standards

In practice, geodesy takes on a number of forms that range from localized surveying conducted by construction and highway crews to three-dimensional positioning at points half a world apart. The accuracy and care required for the latter measurements would be wasted on other applications. For that reason, the United States has established different categories for classification of geodetic control and surveys: **first-order** surveys are the most accurate, **second-order** the next, and **third-order** surveys are the least accurate, although they still meet stringent criteria [5]. Each order may be further subdivided into either **class I** (more accurate) or **class II** (less accurate). Some of the criteria for leveling surveys are listed in Table 3.1. Where ground positions are known with great

TABLE 3.1
Specifications for precise leveling done by the U.S. Coast and Geodetic Survey.

	Order of Leveling			
	First	**Second (Class I)**	**Second (Class II)**	**Third**
Spacing of lines and cross-lines (km)	96.6	40.2–56.3	9.7	Not Specified
Average spacing of permanent benchmarks, not to exceed (km)	1.6	1.6	1.6	4.8
Length of sections (km)	0.8–1.6	0.8–1.6	0.8–1.6	Not Specified
Check between forward and backward runs between fixed elevations, not to exceed (L in km)	4 mm$(L)^{0.5}$	8.4 mm$(L)^{0.5}$	8.4 mm$(L)^{0.5}$	12 mm$(L)^{0.5}$

(After Brown and Oliver, 1976 [8])

certainty, but elevation estimates are less accurate [6], the network is known as a **horizontal geodetic network;** in a **geodetic leveling network,** vertical control points are well established, but the horizontal positions are less accurate.

Measurement of Geodetic Change

For most human purposes, the topography of the Earth is relatively stable through time. At most locations, once position and elevation have been accurately determined, they will not change much during a human lifetime. However, the accuracy of some geodetic measurements is so great that, in tectonically active regions, deformation of the surface of the Earth can be detected over time. All geodetic techniques follow the same basic procedure—a series of measurements precisely determine the relative positions within a network of **control points.**

Resurvey of a geodetic network over a sufficient length of time will lead to an estimate of the rate and direction of motion, if indeed any motion is occurring. The duration of time required to detect neotectonic movement is a function of the *rate* of motion and the *accuracy* of the measurement technique. Along sections of the San Andreas fault, one could measure the northward motion of the Pacific Plate relative to North America with a yardstick, if one were patient enough to wait thirty years or more. With a geodetic positioning technique accurate to ±5 cm combined with a creeping fault that moves steadily at 10 mm/yr, at least 5–10 yr between first and final measurements, and more likely 20–30 yr, are required to say anything meaningful about the motion. With a geodetic positioning technique accurate to ±1 cm, motion on the same fault could be detected by resurveys in as little as 1–2 yr.

Near-Field vs. Far-Field Geodesy

Use of geodesy for tectonic applications is often subdivided into near-field and far-field methods. Far-field geodesy refers to measurement over long distances, away from any one fault in particular. Regional leveling nets are capable of detecting regional tectonic deformation, sometimes alerting scientists to activity that they previously did not know existed [7, 8]. For example, one leg of a regional leveling net crosses the Rio Grande rift in central New Mexico. Between surveys in 1911 and 1951, up to 20 cm of uplift occurred over what is now interpreted as an active magma body beneath the rift [9]. In contrast, near-field geodesy involves measurement in the immediate vicinity, within meters to a few kilometers, of active features or features suspected to be active.

Several far-field techniques are effective in measuring relative movement between points located on different lithospheric plates. The slow motion in the middle of plates seems to be remarkably regular through time [10]. A geodetic survey between London and New York, for example, would detect the continued opening of the Atlantic Ocean at a rate that varies very little over time. Whether or not motion is detected by near-field surveys across a plate boundary or a major fault depends on whether that fault is creeping or locked (see Chapter 1). On creeping faults, the tectonic driving force is accommodated by steady movement along the fault without large earthquakes, movement that can be detected by repeated near-field surveys. On locked faults, the driving force builds up until it exceeds the resistance at the fault plane and an earthquake occurs. A summary of geodetic results in southern California found that constant, creeping motion is occurring on several faults, but that there is no continuous movement on a number of others which are known to be active [11].

GEODETIC TECHNIQUES

The technology used by modern geodesists has changed quite a bit since Eratosthenes and the French Academy of Sciences expeditions. The techniques that represent the current state-of-the-art are informally subdivided as follows:

- ground-based techniques
- space-based techniques
  - Very Long Baseline Interferometry (VLBI)
  - Satellite Laser Ranging (SLR)
  - Global Positioning Systems (GPS)

The ground-based methods are a mature technology, the fundamentals of which have been around for centuries, even millenia. The space-based methods are a new technology, the limitations and applications of which continue to evolve. Each of the different techniques must be evaluated in terms of the equipment required, its potential accuracy and precision, and its possible applications.

Ground-Based Geodesy

Both far-field and near-field geodetic techniques are well suited to studies of neotectonic deformation. More than 700,000 km of leveling lines cross the country, from coasts to mountain peaks, with accuracies measured in centimeters [3]. This control network is sensitive to imperceptible changes in topography such as can be caused by ongoing tectonic activity. For example, in Yellowstone National Park, the geysers and many volcanic features which tourists admire are the result of a shifting magma body in the crust beneath. The area is also seismically active; in 1959, a $M_s = 7.5$ earthquake ruptured and warped the surface just west of the park [12]. A network of geodetic control points, interrelated by high-precision GPS satellite measurements, has been installed across the area in order to understand the year-to-year pattern of surface deformation and how it relates to the driving forces in the crust beneath [13].

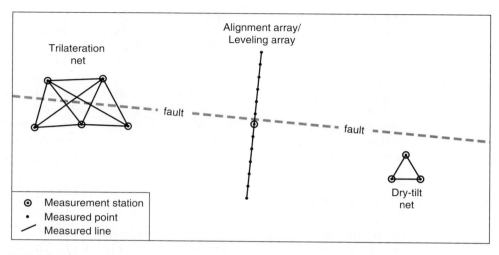

FIGURE 3.4
Near-field geodetic techniques for measuring movement on an active fault. The trilateration net is sensitive to either vertical or horizontal motion, the alignment array to horizontal motion, the leveling array to dip-slip motion, and the dry-tilt net to tilting of the ground surface.

In the immediate vicinity of faults or other geologic structures, near-field geodetic surveying can be used to detect fault creep, uplift or subsidence of the land, tilting, or the growth of active folds [14]. Several common near-field techniques are illustrated in Figure 3.4 and listed below.

- **Alignment or leveling arrays.** Using a single base location, at which a measuring instrument is located, and a line of fixed points which crosses the fault, any relative motion can be determined. Alignment refers to the relative position of the points (sensitive to strike-slip motion), and leveling refers to the elevations (sensitive to vertical motion).
- **Trilateration nets.** A number of stations are located on either side of the fault. Stations are arranged in a pattern of interlocking polygons so that the position of each point is known relative to every other. Repeated measurement of the orientations and lengths of the baselines between the stations results in a network sensitive to minute motion across the fault.
- **Dry-tilt nets** (also called "spirit levels"). These points are generally arranged as an equilateral triangle located on one side of the fault or the other. Any changes in the relative elevations of the points indicate tilting of the ground [14].

Most high-precision surveying today is done with electronic distance measurement (EDM) systems, such as a theodolite (Figure 3.5). These systems bounce a laser pulse off a reflector positioned some distance away and detect

FIGURE 3.5
A theodolite. This instrument and others like it are designed to measure distances and angles with great accuracy. These instruments are the most common devices for measuring geodetic networks (e.g., Figure 3.4).
(Photograph courtesy of Sokkia Corp.)

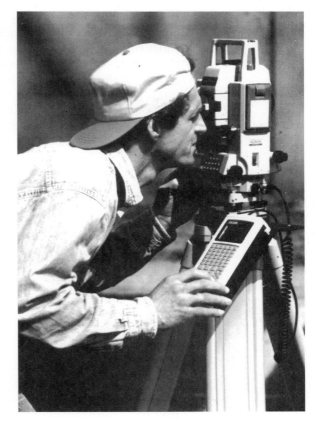

the reflection of the pulse. The time interval between sending the pulse and detecting its reflection is a measure of the distance between the laser source and the reflector. Depending on the equipment used, EDM systems can measure distances up to 10–20 km or more, although greater accuracy is achieved with shorter lines. Under the best conditions, accuracy of measurements over baselines 1–35 km long using a geodolite has been found to be between 3 and 8 mm [15]. This accuracy was achieved by a series of short line sightings, temperature measurements at each station, and humidity measured by overhead aircraft.

Very Long Baseline Interferometry (VLBI)

VLBI staddles the line between geology and astronomy. This technique measures positions on the Earth's surface with accuracy of less than 1 cm by observing quasars, which are among the most distant and most energetic features in the universe. Owing to their great distance from us, quasars are the most stationary beacons in the heavens. Closer celestial objects, such as the stars in the night sky, shift positions at rates detectable by the most sensitive telescopes. Quasars are also desirable as beacons because their radio signals, far higher in frequency than visible light, are virtually undistorted by passage through the Earth's atmosphere [16].

Radio telescopes around the world can track the positions of quasars and monitor the signals they emit. Each receiving station records this information, as well as the exact time at which it was received, calibrated by atomic clocks (Figure 3.6). For any one quasar, the radio signal detected at all of the telescopes is identical, except that the signal will be detected fractionally earlier at

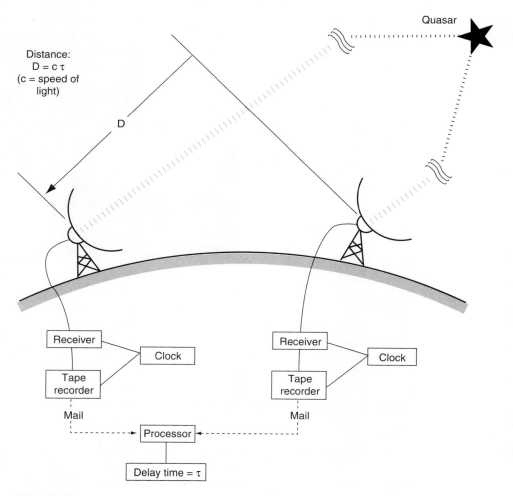

FIGURE 3.6
Distance between two radio telescope sites is measured by recording the signal from a quasar, matching it to atomic clocks, and then comparing the records at the two sites. (After Wells, 1986. *Guide to GPS Positioning*. Canadian GPS Associates: Fredericton, NB)

some stations than at others (Figure 3.7). The lag time between any two receivers is directly related to the distance between them parallel to the quasar's direction:

$$\text{Distance} = \text{rate} \cdot \text{time} = (\text{speed of light}) \cdot (\text{lag time}) \qquad (3.1)$$

By tracking several quasars simultaneously, a network of receiving stations can precisely determine their relative positions in all three dimensions.

The accuracy of these VLBI distance measurements is a few centimeters or less [16]. This method currently is the most accurate of the space-based geodetic techniques. In addition, because VLBI is measured against a celestial reference frame and not a terrestrial one, it provides other information such as precise measurement of the Earth's rotation rate and the orientation of the rotation axis. Very minute variations in the Earth's rotation reflect such processes as tides, winds, seasonal redistribution of water and ice, and flow within the molten interior of the planet [16].

The principal disadvantage of the VLBI technique is that it requires a fully functional radio observatory, equipment that is neither easily portable nor available to the average geodesist. However, monitoring of existing VLBI stations provides a highly accurate, coarsely spaced framework for the shape and the rates of deformation of the surface of the Earth. VLBI measurements have been used successfully in several studies of tectonic movement and seismic deformation [17, 18, 19].

Satellite Laser Ranging (SLR)

In 1976, the United States launched the first fully-dedicated laser geodynamics satellite (LAGEOS). LAGEOS is one of several satellites currently equipped for laser ranging, in which the satellite is used as a reference by which to precisely locate ground positions. Ground stations track the satellite's position in its orbit and bounce laser pulses off it. The distance between the station and the satellite is measured as half of the lag time between sending the laser pulse (t_{send}) and detecting its reflection (t_{detect}), multiplied by the speed of light through the atmosphere (c'):

$$\text{Distance} = 1/2 \cdot (t_{send} - t_{detect}) \cdot c' \qquad (3.2)$$

At the ground station, three types of data are recorded: the distance to the satellite, the satellite's position in the sky above the observing station, and the times of the observations [10, 20] (Figure 3.8).

There are two ways to utilize the data recorded at a SLR ground station. The first is to compare the record with data from another station simultaneously tracking the same laser-ranging satellite. In this method, relative position between the two points can be determined with great accuracy. The second method for locating SLR sites is to use measurements from all stations around the planet over a period of months to calculate a model of the satellite orbit (that model is known as an "**ephemeris**"). The orbits of all satellites are very

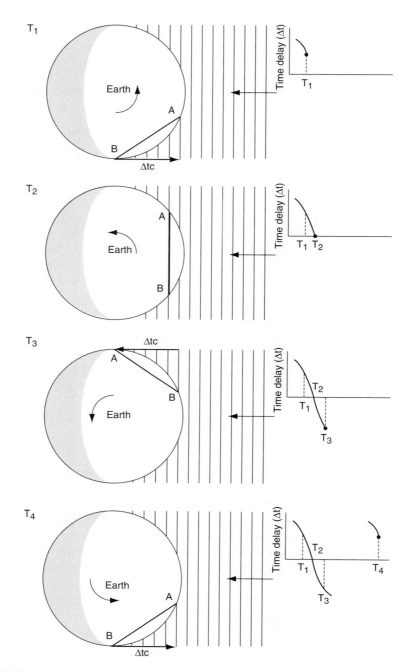

FIGURE 3.7
VLBI measures the arrival times of signals from distant quasars, which can be viewed as a series of parallel wave fronts. Two telescopes on the Earth at points *A* and *B* measure the arrival times *(T₁, T₃)* of a given wave front at the two locations. The duration of the delay *(T₃–T₁)* is used to calculate the distance between the two sites *(Dt • c)*.
(After Carter and Robertson, 1986 [16])

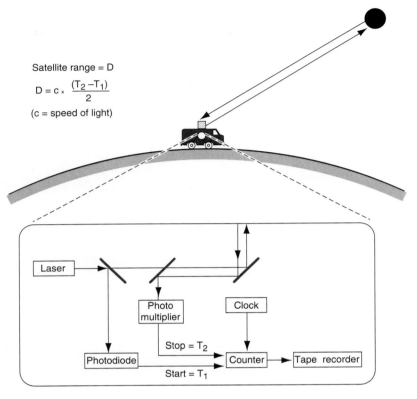

Satellite range = D

$$D = c \times \frac{(T_2 - T_1)}{2}$$

(c = speed of light)

FIGURE 3.8

Satellite Laser Ranging. A laser pulse is sent from a ground station, reflected off the satellite, and returned to the ground. An atomic clock measures the delay between the original signal and its reflection. The range to the satellite is calculated from this delay time. (After Wells, 1986. *Guide to GPS Positioning*. Canadian GPS Associates: Fredericton, NB)

nearly circular or elliptical, but they do follow small variations in the geoid (the equal-gravity shape of the Earth). In addition, orbits are sensitive to factors such as pressure from the Sun's radiation and drag from the highest traces of the atmosphere [20]. The LAGEOS satellite was designed to minimize these orbital variations: it is dense and nearly spherical and is in a high orbit (6000 km) to minimize contact with the atmosphere. The sum of all ground measurements leads to an ephemeris that approximates the satellite's actual position in its orbit to within a few centimeters. A single ground station can determine its own position relative to the model with similar accuracy.

The accuracy of ground locations determined by satellite laser ranging is ±2–3 cm. SLR measurements are made with equipment that is increasingly compact and portable. NASAs Transportable Laser Ranging System (TLRS-2) is packed into a few large crates [20]. A particular weakness of the SLR technique, however, is that it requires favorable weather. Other space-based geodetic methods can operate regardless of weather (barring truly extreme conditions), but a cloudy day renders SLR inoperative.

In some applications of SLR, a very different type of satellite is used to reflect laser pulses back to the geodetic ground station—the Moon. Mirrored reflectors were deployed by the Apollo 11, 14, and 15 missions and by two Soviet Luna landers [21] (Figure 3.9). In Lunar Laser Ranging (LLR), the basic principles

FIGURE 3.9
Lunar Laser Ranging reflector on the Sea of Tranquility, the Moon. This is the first of three such devices emplaced by the Apollo astronauts (note the footprints of Buzz Aldrin in the lunar dust). Laser-ranging facilities on Earth bounce a pulse of light off one of the reflectors and measure the time it takes to detect the reflection. The detectors on Earth must be extremely sensitive because only a few photons will return to their source from a 2-billion watt laser emission.
(From Faller and Dickey, 1990 [21])

of SLR apply. Like SLR geodesy, the position of a ground station can be determined with sufficient precision to measure plate motion. But like VLBI geodesy, LLR ground stations are limited to specially equipped (and nonportable) telescope facilities. Three facilities have been systematically engaged in LLR: McDonald, Texas; Halea Kala, Hawaii; and Grasse, France. Hitting the 46-cm^2 Apollo 11 reflector (Figure 3.9) has been compared with hitting a dime with a rifle fired from two miles away [21], not to mention that you then have to catch the ricochet. LLR reveals new information about the Earth-Moon system; for example, that the drag of tides on Earth is causing the Moon to slip away at a rate of 3.7 cm/yr. Furthermore, LLR has refined our understanding of the Moon's orbit to the point that we can calculate the occurrence of solar eclipses as far back as 1400 B.C. [21, 22]. With some discussion of new manned expeditions to the Moon, there are plans for improved LLR capability that would improve accuracy by nearly two orders of magnitude [23].

Global Positioning System (GPS)

The most promising new geodetic technology is the use of GPS. GPS geodesy uses a constellation of 24 satellites that circle the Earth at an altitude of about 20,000 km and emit signals that are used by receivers on the surface to determine location precisely [24]. The receivers are fully portable; in fact, the most compact designs are hand-held units. Geodesy is just one of the applications of GPS technology; others include marine and aircraft navigation, directing search-and-rescue operations, land surveying, and local deformation modeling (of dams, mines, etc.). With the systematic reduction in the price of GPS equipment, automated maps in cars may someday be standard.

What makes GPS locating possible is the array of satellites operated by the U.S. Department of Defense (Figure 3.10). At the heart of each satellite are four atomic clocks and a radio transmitter, all powered by 7.2 m^2 of solar panels [24]. The GPS satellite array is continuously monitored by a string of permanent stations that gird the globe: Hawaii, Kwajalein, Ascension Island, and Diego Garcia. Information on the satellite orbits, the health of each satellite, and clock accuracies is compiled at Colorado Springs, Colorado, site of the central control station for the system.

Operation of GPS receivers in the field is a passive procedure made possible by the infrastructure that maintains the satellite array (Figure 3.11). Each satellite transmits at two frequencies simultaneously, referred to as L_1 (1575.4 MHz) and L_2 (1227.6 MHz). The signals carry navigation data that is used by receivers on the ground to calculate the distance to the satellite. Receivers also can use the two different frequencies to correct for delays in the signal caused by travel through the atmosphere [25]. Each receiver picks up the signals broadcast by the satellites overhead at that time and uses the distance to each satellite to determine its position on the surface. If three satellites are within range, then the map position of the receiver can be determined; if four or more satellites are in range,

FIGURE 3.10
A GPS satellite. The plan calls for 24 satellites orbiting the Earth, providing 24-hour, three-dimensional positioning capability around the globe.
(Photo by Rockwell/Tsado/Tom Stack & Associates)

then the position and the elevation can be determined. Using a single receiver (**absolute positioning**), the location is accurate to as little as ±10 m [24].

For most applications, accuracy of a few tens of meters is ample, but geodesy requires centimeter- or even millimeter-scale accuracy. With a single receiver, precise positioning is limited by uncertainty in the receiver clock, uncertainties in the ephemeris of each satellite, and slight variations in the speed of the signals through the atmosphere. However, such errors are common to multiple receivers on the ground. By using two or more GPS units (**relative positioning**), with at least one located at a fixed and known point, the positions of other points can be found with accuracies of ±1 cm or less [24]. In fact, with one receiver at a fixed location, a second one, mounted on a vehicle or in a backpack, is free to move around and record positions continuously (**kinematic positioning**). The potential accuracy of GPS measurements has been tested in the very carefully surveyed area along the San Andreas fault near Parkfield,

FIGURE 3.11
A GPS geodetic receiver in operation. Operating in tandem with another receiver on the other side of the bay, the distance between the two sites can be measured to within a few centimeters.
(Photo courtesy of Trimble Navigation)

California. Horizontal coordinates were repeatable to ±4–6 mm and elevations to ±1–2 cm, over baselines up to 11 km long [26]. Furthermore, the uncertainties did not grow proportionally with the distances over which they were measured. Over baselines up to 225 km long, GPS can provide distances accurate to one part in 20 million [26]. Although high-precision surveying with an EDM is still preferable over short baselines, GPS locating is better over longer distances.

GPS geodesy has several advantages over other methods. The equipment works in all weather conditions, it is highly portable and can go wherever the most intrepid geodesist can hike, and receivers working in tandem need not be mutually visible, permitting great flexibility in selecting sites. Its various advantages, especially its potential for great accuracy, make GPS the preferred tool for many applications of geodesy today. For example, repeated GPS surveys are being used to measure the thinning of the Antarctic ice sheet in response to global warming [29], monitoring groundwater withdrawal and land subsidence [30], and pinpointing navigation for offshore oil and gas exploration [31]. A potentially life-saving application of GPS technology is real-time monitoring of volcanic hazards [32]. Volcanic eruptions are often preceded by ground-surface deformation caused by the injection of magma at depth. For example, a network of receivers around Teishi volcano in Japan detected significant deformation of the surface preceding an eruption in July 1989 [33, 34]. A period of increased earthquake activity prior to the eruption was coincident with changes of 4–5 cm in relative elevation and about 13 cm in distance between two stations on either side of the volcanic center (Figure 3.12). This monitoring has the potential to be fully automated; receivers permanently deployed at trouble spots could alert scientists to impending danger.

It is in the area of active tectonics, however, that GPS may have its greatest scientific value. Although the technology has been available for only a few years, GPS has been used to measure rates of interplate movement and crustal deformation around the world [35, 36, 37, 38]. In addition, the pattern of coseismic surface deformation that accompanies a major earthquake can help reveal the geometry of faulting, even where the fault doesn't rupture the surface [39]. GPS measurements have the potential to detect movement before, during, and after earthquakes (see discussion of the seismic deformation cycle). Rapid increases in the number of GPS monitoring sites and the speed at which GPS data can be processed suggest that the technique may someday contribute to prediction *in advance* of earthquakes.

GPS technology is still in its infancy. Although the number of receivers in use is increasing rapidly, several obstacles remain that block an equal increase in scientific results. It has been suggested that, in some cases, the precision of GPS measurements has outstripped the stability of the benchmarks being measured [40]. Benchmarks should be in solid rock; among the best are stainless steel pins drilled into bedrock. Another obstacle is that geodetic-quality GPS measurements require extensive and time-consuming analysis before the results are in hand. The ability to collect GPS information may have outstripped the ability to process and make sense of it [40].

A final obstacle to widespread use of GPS geodesy is military security. The GPS satellite constellation was designed and is operated principally as a military tool. Until recently, GPS satellites have broadcast their signals free and without restrictions. Since 1990, however, the U.S. Department of Defense has considered it necessary to intentionally degrade the GPS signal [41]. The military has two techniques for restricting access to the most precise GPS-positioning capability: Selective Availability and Anti-Spoofing [25]. Selective Availability involves degrading the broadcast information about satellite orbits and perturbing the satellite clocks. Anti-Spoofing refers to encryption of the L_2 signal. The highest possible precision is still available to receivers equipped with appropriate decoding devices. There is a possibility that Selective Availability decoders may be provided to scientific users who require the maximum possible accuracy.

Summary of Geodetic Techniques

The utility and applicability of each of the techniques discussed here depends on the equipment required, the accuracy achieved, and the distances over which

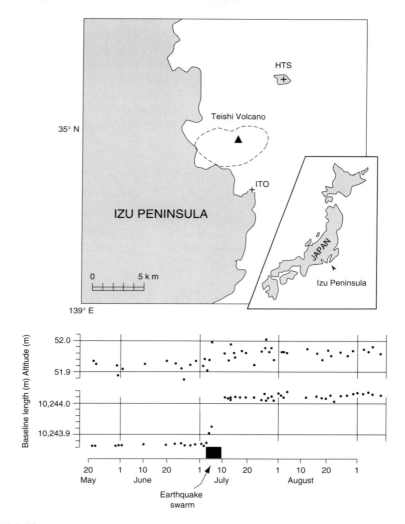

FIGURE 3.12
GPS monitoring of Teishi volcano, Japan. A swarm of earthquakes between July 3 and 10 preceded the eruption of July 12, 1989. GPS receivers at two locations across the volcanic feature, ITO and HTS, detected permanent changes in relative elevation and distance during the increased seismic activity.
(After Shimada et al., 1990 [33])

that accuracy can be obtained. Figure 3.13 illustrates accuracy of relative positioning as a function of baseline length. Ground-based measurement (using an EDM) remains the most accurate method for all baseline distances less than about 4 km. Of the spaced-based technologies, VLBI is the most accurate, although not necessarily the most useful, given the great mass and complexity of equipment required. The flattest line on Figure 3.13 describes the accuracy of

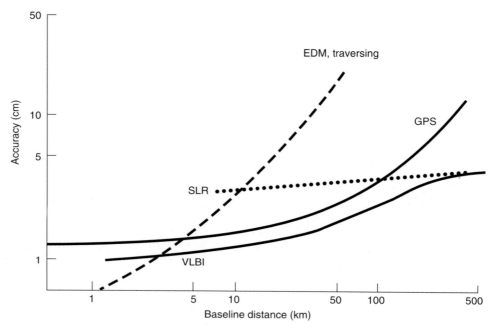

FIGURE 3.13
Accuracy as a function of baseline length for a variety of geodetic techniques. Electronic
distance measurement, for example, is most accurate over baselines up to a few kilometers in length, but becomes decreasingly reliable (relative to the other techniques) over
longer distances.
(After Wells, 1986. *Guide to GPS Positioning.* Canadian GPS Associates: Fredericton, NB)

SLR, which fades little with increasing distance. However, the accuracy of the
SLR technique may be counterbalanced by the cost of the equipment and its
limitations in nonideal weather conditions. Probably the best balance of accuracy, cost, and mobility today is GPS, which may well become the standard for
many geodetic and other applications in the near future.

APPLICATION OF GEODESY TO ACTIVE TECTONICS

Seismic Deformation Cycle

Geodetic measurements can reveal many details of neotectonic motion. Comparison of far-field and near-field results tells us much about how stress builds up at
faults and leads to earthquakes. Where far-field measurements detect motion, it
usually represents steady, long-term movement, such as the inexorable motion
between two plates. Where the near-field measurements across a given plate
boundary equal the far-field motion, we can conclude that the fault is **creeping**

(see Chapter 1) and that motion is concentrated on that fault strand spanned by the near-field array. Recall that creep is defined as any motion on a fault that is not accompanied by measurable seismic activity.

In many cases, however, active faults remain frozen, building up stress over decades to centuries during which no near-field deformation is observed (see Chapter 1). On such locked faults, most or all movement is **coseismic motion,** instantaneous and coincident with an earthquake. When a fault ruptures the surface, it typically produces deformation that is quite unsubtle. However, a large earthquake can cause deformation both near the fault zone and over large surrounding areas. The first large earthquake ever monitored by GPS geodesy was the magnitude 7.5 Landers earthquake, which shook most Southern Californians out of their beds at 4:57 a.m. on June 28, 1992. The Landers earthquake was the largest in California in 40 years and was over six times as powerful as the 1994 Northridge earthquake which caused an estimated $20 billion in damage in the Los Angeles area. Damage and loss of life due to the Landers event were minimal because that earthquake was centered in a remote area northeast of Los Angeles [42]. The Permanent GPS Geodetic Array (PGGA) was established in Southern California in 1991, just in time to measure the effects of the Landers earthquake. The earthquake caused surface rupture with displacements up to 5 m along a 85-km-long zone of the Mojave Desert. The GPS stations also revealed something that was not immediately obvious, that the earthquake altered the shape of Southern California—shifting the land by up to 0.5 m as far as 100 km from the fault zone [43, 44].

Regional deformation associated with the Landers earthquake was also detected and measured by a new geodetic technique, **satellite radar interferometry** (Figure 3.14). Some satellites are equipped with synthetic aperture radar, which measures the distance between the satellite and the Earth's surface by bouncing radar signals off the surface at various locations in the satellite's orbit. These satellites collect data continuously, repeatedly mapping regions at regular intervals. A radar-mapping satellite, ERS-1, passed over the Mojave Desert before and after the 1992 Landers earthquake. Changes in the satellite-to-surface distance near the region of faulting reveal a systematic pattern called an **interferogram** (Figure 3.14), which is a regional map of deformation caused by the earthquake. This technique is superior to GPS measurements because it creates a picture of movement across the entire area, rather than measurements at a small number of points [45].

Deformation associated with the Landers earthquake seems to have occurred largely in a single coseismic pulse [44], but other faults have been known to exhibit creeping motion immediately before and/or after an earthquake. **Afterslip,** which is aseismic motion that follows in the days or few weeks after a coseismic rupture, may be relatively rapid immediately after the earthquake, but usually diminishes to zero with time [14]. An earthquake rupture also may be preceded by **preseismic motion.** Using near-field geodesy to detect preseismic motion is of particular interest as a means of predicting earthquakes in

FIGURE 3.14
Radar interferogram illustrating coseismic motion that accompanied the Landers earthquake sequence of June 28, 1992. Each dark band represents 28 mm of change in position, so that the image represents a contour map of deformation. In this image, the original radar data were used to create a geophysical model of coseismic deformation that smooths the contours and eliminates some of the random noise in the signal. (From Massonnet et al., 1993 [45])

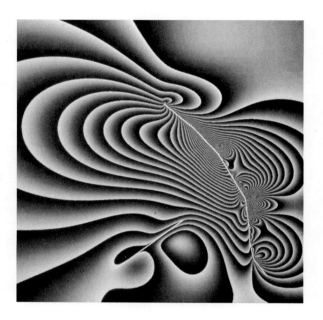

the days before they occur. Altogether, the repeated pattern of coseismic motion during an earthquake, postseismic motion, an interseismic interval, preseismic motion, and another earthquake is known as the **seismic deformation cycle** [46] (Figure 3.15).

Geologic Models of Plate Motion

All geodetic measurements of tectonic movement are the briefest snapshots of motions that have been occurring for long periods of geologic time. The geodetic estimates of present plate motion need to be compared to estimates from longer intervals. Several aspects of the geology of the plates and the plate boundaries preserve the directions and rates of motion [9]:

- oceanic magnetic anomalies
- orientations of ocean transform faults
- offsets of geologic formations and features

Earthquake focal mechanisms—the sense of rupture measured during earthquakes by seismographs around the world—also help delineate the plates and show how they are moving. All of this geologic information has been incorporated into global models of lithospheric plate motion [47].

FIGURE 3.15
Schematic illustration of the seismic deformation cycle. (A) Cycle with no permanent deformation—interseismic motion equals coseismic slip. (B) Cycle with permanent deformation in the same direction as the coseismic movement—the sum of the postseismic and interseismic motion is less than the coseismic. (C) Cycle with permanent deformation of the opposite sense as the coseismic movement—postseismic and interseismic exceeds the coseismic.
(After Thatcher, 1986 [46])

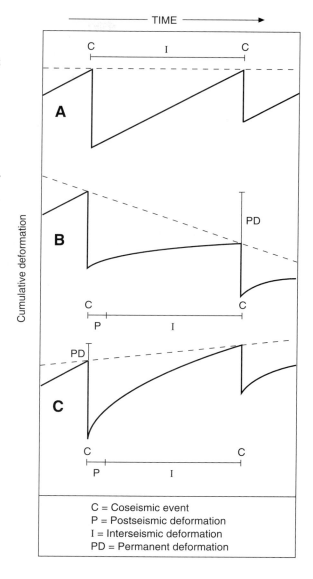

C = Coseismic event
P = Postseismic deformation
I = Interseismic deformation
PD = Permanent deformation

Comparing the various geodetic measurements with the geologic record of plate motion, the most striking conclusion is how very close the estimates are [10, 48]. This is a testament not only to the accuracy of the new geodetic technology, but also to the remarkable regularity of plate motion. We must conclude that, although plates may stick and slip at their edges, the great bulk of the lithospheric plates rolls smoothly along with little or no variation in velocity over millions of years [48].

CASE STUDY

"MISSING MOTION" ON THE SAN ANDREAS FAULT

Despite generally close agreement between geodetic measurements and geologic information on the San Andreas fault, some notable discrepancies exist. For example, the far-field rate of motion between the Pacific and North American plates is about 48 mm/yr [49], but the rate of motion on the San Andreas fault in central California is only 34 ± 2 mm/yr [50]. The San Andreas fault is mapped as the boundary between the two plates, but either it has slipped unusually slowly in recent times or, more likely, this *one* fault alone does not accommodate *all* of the motion that we know is occurring. The problem of this "missing motion" is summarized in Figure 3.16, which illustrates both the rates and directions of movement in the form of vectors.

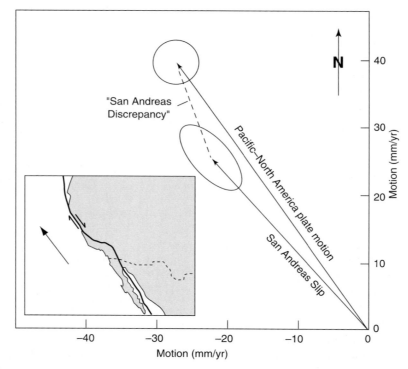

FIGURE 3.16

The San Andreas discrepancy. Both the relative motion between the Pacific and North American plates and slip on the San Andreas fault are displayed as vectors, illustrating directions and magnitudes of motion (including error ellipses). The "missing motion" on the San Andreas is represented by the gap between the two vectors, representing motion that must be accounted for on faults other than the San Andreas.
(After Jordan and Minster, 1988 [51])

FIGURE 3.17

Simplified tectonic map of the western United States showing the Pacific plate, the Sierra Nevada–Great Valley block (SN), the Great Basin region (GB), and the stable portion of the North American plate. Solid squares are radio telescope sites on North America; triangles are radio telescopes on the Sierra Nevada block.

(After Argus and Gordon, 1991 [52])

Two simplifying assumptions incorporated into most current global plate models are:

(1) the plates are rigid bodies, with no interior deformation, and

(2) plate boundaries are sharp, concentrated features where all relative plate motion takes place.

The plate models fail where the tectonic reality is not so simple, where (1) some deformation occurs in the interior of plates (**intraplate motion**), or (2) where the plate boundaries are not simple lines on the map, but rather broad zones across which relative motion is distributed [48]. In the case of the San Andreas, both simplifying assumptions seem to be invalid.

West of the San Andreas fault, on the North American landmass but on the Pacific plate (Figure 3.17), the Coast Ranges and the Continental Borderland are cut by faults and folds that accommodate compression and right-lateral translation. East of the San Andreas fault, the Great Valley and Sierra Nevada appear to constitute a rigid, undeformed block. However, the Sierra Nevada is separated from the core of stable North America by the Great Basin province (also called the Basin and Range), a broad zone of active intraplate deformation. The Great Basin is characterized by many deep basins and steep, fault-bounded mountain blocks. This topography was formed by widespread extension and right-lateral translation during the past 20–30 m.y. It is theoretically possible, but unlikely in practice, to find the total rate of deformation across the Great Basin by purely geological means—studying every fault in the province and summing the total displacement.

The great strength of space-based geodesy is its ability to determine the relative motion of *any* two points on the surface, not only between great ocean basins or across simple, concentrated bounding faults. Radio telescopes on both the west and the east sides of the Great Basin allow us to measure the intraplate motion east of the San Andreas fault using VLBI [51]—the result, relative movement of 11 ±1 mm/yr is oriented 36° ± 3° west of north [52]. When this vector is fit into the "missing motion," it accounts for all

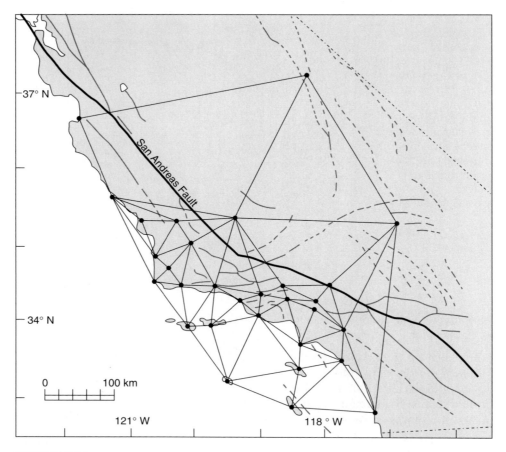

FIGURE 3.18
A geodetic network monitors deformation west of the San Andreas fault, across the California Coast Ranges and Continental Borderland. Because motion between the Pacific Plate and the North American Plate is not limited to the San Andreas fault, but is distributed over a broad range of faults, this dense network was designed to determine exactly where this motion is occurring.
(After Jordan and Minster, 1988 [51])

but 6 ± 2 mm/yr of the San Andreas discrepancy. That remainder is probably taken up by deformation west of the fault. The final step in accounting for all of the motion between the Pacific and North American plates is to measure where on the California coast those remaining 6 mm/yr are going. A GPS grid is being monitored to measure deformation across the California Coast Ranges and Continental Borderland to help determine where the remainder of the motion is occurring [51] (Figure 3.18).

SUMMARY

Rapid advances in the technology of high-precision geodesy are leading to a revolution in the study of active tectonics, global geology, and astronomy, as well as countless nonscientific applications. In active tectonics, it is possible for the first time not only to reconstruct surface deformation, but actually to measure it directly! Repeated geodetic surveys can detect the directions and rates at which lithospheric plates move. The relative motion of the centers of the plates is remarkably regular, but stress and strain are unevenly distributed near the plate boundaries. At more local scales, geodetic networks are being used to detect regional warping of the surface and creep on faults. As technology and data-processing capability improve, it is likely that geodesy increasingly will be used to provide real-time warning of volcanic and earthquake hazards.

REFERENCES CITED

1. Torge, W., 1980 [translated by C. Jekeli]. *Geodesy: An Introduction.* Walter de Gruyter: New York.

2. Vanícek, P., and E.J. Krakiwsky, 1986. *Geodesy: The Concepts.* North-Holland: New York.

3. Fernie, J.D., 1991 and 1992. The shape of the Earth, Parts I, II, and III. *American Scientist,* 79: 108–110, 393–395; 80: 125–127.

4. Emery, K.O., and D.G. Aubrey, 1991. *Sea Levels, Land Levels, and Tide Gauges.* Springer-Verlag: New York.

5. U.S. Federal Geodetic Control Committee, 1984. *Standards and Specifications for Geodetic Control Networks.* National Oceanographic and Atmospheric Administration: Rockville, MD.

6. Schomaker, M.C., and R.M. Berry, 1981. *Geodetic Leveling.* NOAA Manual NOS NGS 3, U.S. Dept. of Commerce: Washington, D.C.

7. Fujita, N., 1974. Horizontal displacements in Japan. *Tectonophysics,* 23: 419–422.

8. Brown, L.D., and J.E. Oliver, 1976. Vertical crustal movements from leveling data and their relation to geologic structure in the eastern United States. *Reviews of Geophysics and Space Physics,* 14: 13–35.

9. Reilinger, R., and J. Oliver, 1976. Modern uplift associated with a proposed magma body in the vicinity of Socorro, New Mexico. *Geology,* 4: 583–586.

10. Harrison, C.G.A., and N.B. Douglas, 1990. Satellite laser ranging and geological constraints on plate motion. *Tectonics,* 9: 935–952.

11. Louie, J.N., C.R. Allen, D.C. Johnson, P.C. Haase, and S.N. Cohn, 1985. Fault slip in southern California. *Bulletin of the Seismological Society of America,* 75: 811–833.

12. Witkind, I.J., W.B. Myers, J.B. Hadley, W. Hamilton, and G.D. Fraser, 1962. Geologic features of the earthquake at Hebgen Lake, Montana, August 17, 1959. *Bulletin of the Seismological Society of America,* 52: 163–180.

13. Meertens, C.M., and R.B. Smith, 1991. Crustal deformation of the Yellowstone caldera from first GPS measurements: 1987–1989. *Geophysical Research Letters,* 18: 1763–1766.

14. Sylvester, A.G., 1986. Near-field tectonic geodesy. In *Active Tectonics,* National Academy Press: Washington D.C.

15. Savage, J.C., and W.H. Prescott, 1973. Precision of Geodolite distance measurements for determining fault movements. *Journal of Geophysical Research,* 78: 6001–6008.

16. Carter, W.E., and D.S. Robertson, 1986. Studying the Earth by very-long-baseline interferometry. *Scientific American,* 255 (5): 46–54.

17. Molnar, P., and J.M. Gibson, 1994. Very long baseline interferometry and active rotations of crustal blocks in the Western Transverse Ranges, California. *Geological Society of America Bulletin,* 106: 594–606.

18. Zarraoa, N., A. Rius, E. Sardon, and J.W. Ryan, 1994. Relative motions in Europe studied with a geodetic VLBI network. *Geophysical Journal International,* 117: 763–768.

19. Argus, D.F., and G.A. Lyzenga, 1994. Site velocities before and after the Loma Prieta and Gulf of Alaska earthquakes determined from VLBI. *Geophysical Research Letters,* 21: 333–336.

20. Christodoulidis, D.C., D.E. Smith, R. Kolenkiewicz, S.M. Klosko, M.H. Torrence, and P.J. Dunn, 1985. Observing tectonic plate motions and deformations from satellite laser ranging. *Journal of Geophysical Research,* 90: 9249–9263.

21. Faller, J.E., and J.O. Dickey, 1990. Lunar laser ranging. *EOS: Transactions, American Geophysical Union,* 71: 725–726.

22. Morrison, D.C., 1989. An unsung legacy of the first lunar landing. *Science,* 246: 447–448.

23. Bender, P.L., and J.E. Faller, 1990. Ranging goals for our return to the Moon. *EOS: Transactions, American Geophysical Union,* 71: 475 (abstract G21A-9).

24. Leick, A., 1990. *GPS Satellite Surveying.* John Wiley & Sons: New York.

25. Prescott, W.H., 1994, personal communication.

26. Prescott, W.H., J.L. Davis, and J.L. Svarc, 1989. Global positioning system measurements for crustal deformation: Precision and accuracy. *Science,* 244: 1337–1340.

27. Vega, V., J. Kellogg, and J.T. Freymueller, 1992. Repeat observations over northwestern South America, the Caribbean and Panama from the CASA Global Positioning System project. *EOS: Transactions, American Geophysical Union,* 73: 86 (abstract G42A-7).

28. Dixon, T.H., S.K. Wolf, G. Blewitt, and M. Heflin, 1992. Preliminary results from the CASA GPS experiment. *EOS: Transac-*

tions, American Geophysical Union, 73: 86 (abstract G42A-6).

29. Shibuya, K., Y. Fukuda, and Y. Michida, 1990. Applications of GPS relative positioning for height above sea level in the Antarctic marginal ice zone. *Journal of Physics of the Earth,* 38: 149–162.

30. Blodgett, J.C., M.E. Ikehara, and G.E. Williams, 1990. Monitoring land subsidence in Sacramento Valley, California, using GPS. *Journal of Surveying Engineering,* 116: 112–130.

31. Jensen, M.H.B., 1992. GPS in offshore oil and gas exploration. *Geophysics: The Leading Edge of Exploration,* 11 (11): 30–34.

32. Thatcher, W., 1990. Precursors to eruption. *Nature,* 343: 590–591.

33. Shimada, S., Y. Fujinawa, S. Sekiguchi, S. Ohmi, T. Eguchi, and Y. Okada, 1990. Detection of a volcanic fracture opening in Japan using Global Positioning System measurements. *Nature,* 343: 631–633.

34. Yabuki, T., T. Kanazawa, and H. Wakita, 1991. Anomalous movements in Oshima volcano associated with the Ito submarine eruption revealed from GPS measurements. In Y. Ida and M. Mizoue (eds.), Seismic and Volcanic Activity In and Around the Izu Peninsula and its Tectonic Implications. *Journal of Physics of the Earth,* 39: 155–164.

35. Williams, S.D.P., J.L. Svarc, M. Lisowski, and W.H. Prescott, 1994. GPS measured rates of deformation in the northern San Francisco Bay region, California, 1990–1993. *Geophysical Research Letters,* 21: 1511–1514.

36. Sturkell, E., F. Sigmundsson, P. Einarsson, and R. Bilham, 1994. Strain accumulation 1986–1992 across the Reykjanes Peninsula plate boundary, Iceland, determined from GPS measurements. *Geophysical Research Letters,* 21: 125–128.

37. Dixon, T.H., 1993. GPS measurement of relative motion of the Cocos and Caribbean Plates and strain accumulation across the Middle America Trench. *Geophysical Research Letters,* 20: 2167–2170.

38. Tanaka, T., and M. Ohba, 1993. Crustal movement research and GPS in Japan. *Journal of Geodynamics,* 18: 1–12.

39. Williams, C.R., T. Arnadottir, and P. Segall, 1993. Coseismic deformation and dislocation models of the 1989 Loma Prieta earthquake derived from global positioning system measurements. *Journal of Geophysical Research,* 98: 4567–4578.

40. Bevis, M., 1991. GPS networks: the practical side. *EOS: Transactions, American Geophysical Union,* 72: 49, 55–56.

41. Anderson, G.C., and D. Swinbanks, 1990. Scientists given the jitters. *Nature,* 345: 195.

42. Hudnut, M.J., L. Jones, E. Hauksson, and K. Hutton, 1992. Rapid scientific response to Landers quake. *EOS: Transactions, American Geophysical Union,* 73: 417–418.

43. Bock, Y., D.C. Agnew, P. Fang, J.F. Genrich, B.H. Hager, T.A. Herring, K.W. Hudnut, R.W. King, S. Larsen, J.B. Minster, K. Stark, S. Wdowinski, and F.K. Wyatt, 1993. Detection of crustal deformation from the Landers earthquake sequence using continuous geodetic measurements. *Nature,* 361: 337–340.

44. Blewitt, G., M.B. Heflin, K.J. Hurst, D.C. Jefferson, F.H. Webb, and J.F. Zumberge, 1993. Absolute far-field displacements from the 28 June 1992 Landers earthquake sequence. *Nature,* 361: 340–342.

45. Massonnet, D., M. Rossi, C. Carmona, F. Adragna, G. Peltzer, K. Feigl, and T. Rabaute, 1993. The displacement field of the Landers earthquake mapped by radar interferometry. *Nature,* 364: 138–142.

46. Thatcher, W., 1986. Geodetic measurements of active-tectonic processes. In *Active Tectonics.* National Academy Press: Washington, D.C.

47. DeMets, C., R.G. Gordon, D.F. Argus, and S. Stein, 1990. Current plate motions. *Geophysical Journal International,* 101: 425–478.

48. Gordon, R.G, and S. Stein, 1992. Global tectonics and space geodesy. *Science,* 256: 333–342.

49. DeMets, C., R.G. Gordon, S. Stein, and D.F. Argus, 1987. A revised estimate of Pacific–North America motion and implications for western North American plate boundary zone tectonics. *Geophysical Research Letters,* 14: 911–914.

50. Sieh, K.E., and R.H. Jahns, 1984. Holocene activity of the San Andreas fault at Wallace Creek, California. *Geological Society of America Bulletin,* 95: 883–896.

51. Jordan, T.H., and J.B. Minster, 1988. Measuring crustal deformation in the American West. *Scientific American,* 259 (2): 48–56.

52. Argus, D.F., and R.G. Gordon, 1991. Current Sierra Nevada–North America motion from very long baseline interferometry: implications for the kinematics of the western United States. *Geology,* 19: 1085–1088.

4
Geomorphic Indices of Active Tectonics

INTRODUCTION

Morphometry is defined as quantitative measurement of landscape shape. At the simplest level, landforms can be characterized in terms of their size, elevation (maximum, minimum, or average), and slope. Quantitative measurements allow geomorphologists to objectively compare different landforms and to calculate less straightforward parameters (**geomorphic indices**) that may be useful for identifying a particular characteristic of an area—for example, its level of tectonic activity.

Some geomorphic indices have been developed as basic reconnaissance tools to identify areas experiencing rapid tectonic deformation [1]. This information is used for planning research to obtain detailed information about active tectonics. Other indices were developed to quantify description of landscape [2]. Geomorphic indices are particularly useful in tectonic studies because they can be used for rapid evaluation of large areas, and the necessary data often are obtained easily from topographic maps and aerial photographs [1]. Some of the geomorphic indices most useful in studies of active tectonics are:

- the hypsometric integral [2]
- drainage basin asymmetry [3, 4]
- stream length–gradient index [5]
- mountain front sinuosity (S_{mf} index) [6, 7]
- ratio of valley floor width to valley height (V_f index) [6, 7].

The results of several of the indices may also be combined, along with other information such as uplift rates, to produce **tectonic activity classes** [7], which are broad-based assessments of the relative degree of activity in an area.

HYPSOMETRIC CURVE AND HYPSOMETRIC INTEGRAL

The **hypsometric curve** describes the distribution of elevations across an area of land, from one drainage basin to the entire planet. The curve is created by plotting the proportion of total basin height (relative height) against the proportion of total basin area (relative area) (Figure 4.1) [2]. A hypsometric curve for a drainage basin on a uniform slope (Figure 4.2) illustrates how the curve is created. The drainage basin spans eight contour lines, numbered 1 to 8 on the figure. The total surface area of the basin *(A)* is the sum of the area between each pair of adjacent contour lines. The area *(a)* is the surface area within the basin above a given line of elevation *(h)*. The value of relative area *(a/A)* always varies from 1.0 at the lowest point in the basin *(h/H = 0.0)* to 0.0 at the highest point in the basin *(h/H = 1.0)*.

A useful attribute of the hypsometric curve is that drainage basins of different sizes can be compared with each other because area and elevation are plotted as functions of total area and total elevation. That is, the hypsometric curve is independent of differences in basin size and relief [2]. As long as the topographic maps being used are of a sufficiently large scale to accurately characterize the basins being measured, there should be no effect of different scales.

A simple way to characterize the shape of the hypsometric curve for a given drainage basin is to calculate its **hypsometric integral.** The integral is

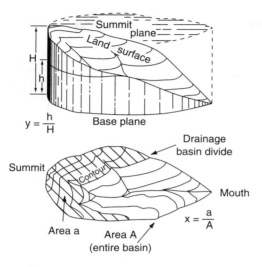
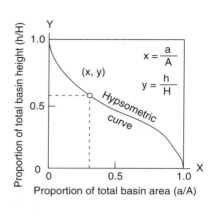

FIGURE 4.1

Hypothetical drainage basin showing how one point (x,y) on the hypsometric curve is derived. Plotting several other values (for different contours) of *a/A* and *h/H* allows the curve to be constructed.

(After Strahler, 1952 [2])

defined as the area under the hypsometric curve. One way to calculate the integral for a given curve is as follows [8, 9]:

$$\frac{\text{mean elevation} - \text{minimum elevation}}{\text{maximum elevation} - \text{minimum elevation}} \qquad (4.1)$$

Thus only three values, easily obtained from a topographic map, are necessary to calculate the integral. Maximum and minimum elevations are read directly from the map. Mean elevation is obtained by point sampling (on a grid) of at least 50 values of elevation in the basin and calculating the mean [8], or by using Digital Elevation Models (DEMs). High values of the hypsometric integral indicate that most of the topography is high relative to the mean, such as a smooth upland surface cut by deeply incised streams. Intermediate to low values of the integral are associated with more evenly dissected drainage basins.

The relationship between the hypsometric integral and degree of dissection permits its use as an indicator of a landscape's stage in the cycle of erosion (see Chapter 2). The cycle of erosion describes the theoretical evolution of a landscape through several stages: a "youthful" stage characterized by deep incision and rugged relief, a "mature" stage where many geomorphic processes operate in approximate

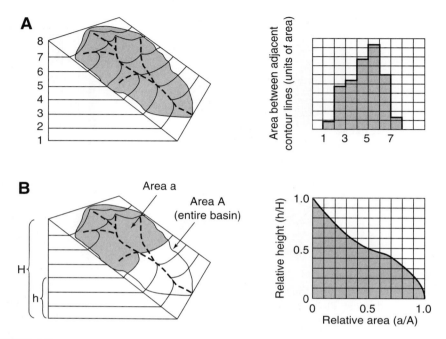

FIGURE 4.2
Idealized diagram showing how the hypsometric curve is determined. See text for further explanation.
(After Mayer, 1990 [9])

equilibrium, and an "old age" stage characterized by a landscape near base level with very subdued relief. A high hypsometric integral indicates a youthful topography (Figure 4.3A). An intermediate value of the hypsometric integral and a sigmoidal-shaped hypsometric curve indicate a mature stage of development (Figure 4.3B). Further development to the old-age stage will not change the value of the integral, unless high-standing erosional remnants are preserved (Figure 4.3C). However, more sophisticated numerical descriptions of the hypsometric curve that are sensitive to continued evolution of the topography are available [2, 9, 10]. In summary, hypsometric analysis remains a powerful tool for differentiating tectonically active from inactive regions. The calculation of hypsometric curves and integrals has become almost trivial with the advent of digital elevation models (DEMs) [11]. Hypsometry at continental and planetary scales is discussed in Chapter 9.

DRAINAGE BASIN ASYMMETRY

The geometry of stream networks can be described in several ways, both qualitatively and quantitatively. Where drainage develops in the presence of active

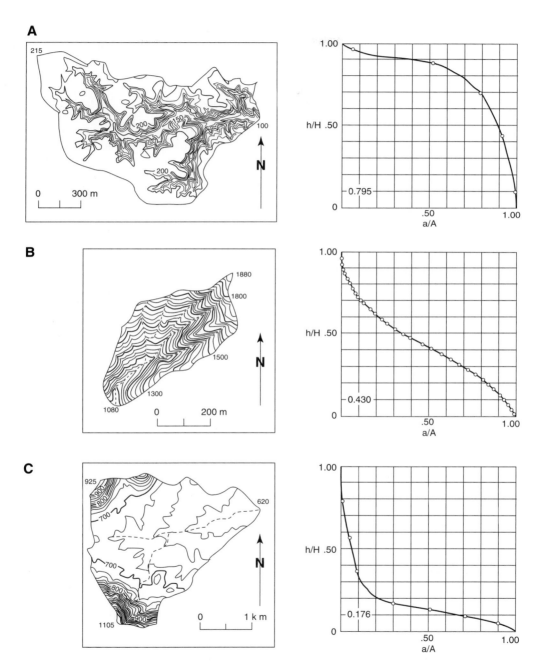

FIGURE 4.3
Three examples of different values of the hypsometric integral. See text for explanation.
(After Strahler, 1952 [2])

tectonic deformation, the network often has a distinct pattern and geometry (see Chapter 6). The **Asymmetry Factor** was developed to detect tectonic tilting at drainage-basin scales or larger areas [3].

The Asymmetry Factor *(AF)* is defined as:

$$AF = 100 \ (A_r / A_t) \tag{4.2}$$

where A_r is the area of the basin to the right (facing downstream) of the trunk stream, and A_t is the total area of the drainage basin. For a stream network that formed and continues to flow in a stable setting, AF should equal about 50. The AF is sensitive to tilting *perpendicular* to the trend of the trunk stream. Values of AF greater or less than 50 may suggest tilt. For example, in a drainage basin where the trunk stream flows north and tectonic rotation is down to the west (Figure 4.4), tributaries on the east (right) side of the main stream are long compared to tributaries on the west side, and AF is greater than 50. If the tilting was in the opposite direction, then the larger streams would be on the left side of the main stream and the AF would be less than 50.

Like most geomorphic indices, the AF works best where each drainage basin is underlain by the same rock type. The method also assumes that neither lithologic controls (such as dipping sedimentary layers) nor localized climate (such as vegetation differences beyween north- and south-facing slopes) causes the asymmetry [12].

FIGURE 4.4
Block diagram showing how the asymmetry factor is calculated.

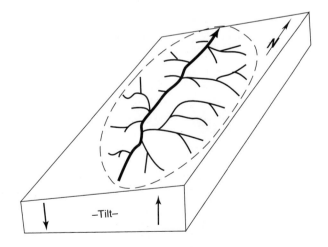

$$AF = 100 \left(\frac{A_r}{A_t} \right)$$

$$= 100 \left(\frac{3.2 \ km^2}{4.9 \ km^2} \right) = 65$$

AF > 50 implies tilt down to the left of basin (looking downstream)

An example of application of the drainage basin asymmetry factor comes from the Pacific coast of Costa Rica [3]. The Nicoya Peninsula is a broad, emergent area of the outer arc of the Middle America subduction zone. The Nicoya Peninsula has been the site of active uplift probably since at least Oligocene–Miocene time (about 25 Ma). Measurements of the drainage-basin asymmetry of large southwest-draining rivers show that deformation coincides with the major faults on the peninsula (Figure 4.5). In particular, the area southeast of the Montaña Lineament Zone is marked by drainage basins tilted down to the southeast. Together, the morphometric and the geologic data support a faulted half-dome model for the active deformation of the Nicoya Peninsula [3].

Another quantitative index to evaluate basin asymmetry is the **Transverse Topographic Symmetry Factor** *(T)* [4]

$$T = D_a / D_d \qquad (4.3)$$

where D_a is the distance from the midline of the drainage basin to the midline of the active meander belt, and D_d is the distance from the basin midline to the basin divide (Figure 4.6). For a perfectly symmetric basin, $T = 0$. As asymmetry increases, T increases and approaches a value of 1. Assuming that the dip of the bedrock can be shown to have negligible influence on the migration of stream channels, then the direction of regional migration is an indication of the ground tilting in that direction [4]. Thus, T is a vector with a bearing (direction) and magnitude from 0 to 1. Values of T are calculated for different segments of valleys (Figure 4.6) and indicate preferred migration of streams perpendicular to the

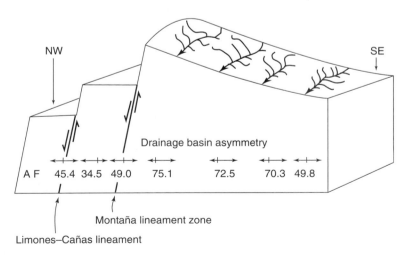

FIGURE 4.5
Asymmetry factors for the Nicoya Peninsula indicate tilting down to the southeast, south of the Montaña Lineament.
(After Hare and Gardner, 1985 [3])

FIGURE 4.6
Diagram of a portion of a
drainage basin showing how
the transverse topographic
symmetry factor *(T)* is
calculated.
(After Cox, 1994 [4])

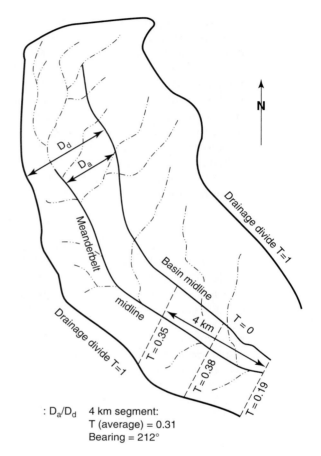

: D_a/D_d 4 km segment:
T (average) = 0.31
Bearing = 212°

drainage-basin axis. This analysis is most appropriate to dendritic drainage patterns, where evaluation of tributary valleys as well as the main or trunk valley allows for a larger range of *T*. Statistical analysis can then be used to estimate most prominent direction of stream migration. This method, as with *AF* described earlier, does not provide direct evidence of ground tilting, but like *AF*, it is a method for rapidly identifying possible tilt. This method was used to suggest the direction of possible Holocene tilting of the ground in the Mississippi Embayment [4].

STREAM LENGTH–GRADIENT INDEX

The **stream length–gradient index** (or *SL* index) is calculated for a particular reach of interest and defined as:

$$SL = (\Delta H / \Delta L) \bullet L \qquad (4.4)$$

FIGURE 4.7
Idealized diagram showing how stream length–gradient index *(SL)* is calculated for the hypothetical "Hack Creek."

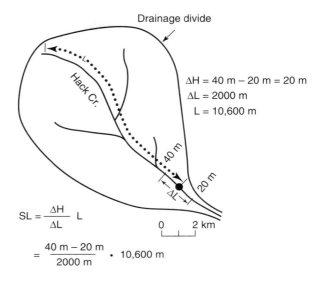

where *SL* is the stream length–gradient index, *ΔH/ΔL* is the channel slope or gradient of the reach (*ΔH* is the change in elevation of the reach and *ΔL* is the length of the reach), and *L* is the total channel length from the point of interest where the index is being calculated upstream to the highest point on the channel. In most cases, these parameters are measured from topographic maps. Figure 4.7 shows how the *SL* index is calculated for a hypothetical example.

The *SL* index correlates to stream power. Total stream power available at a particular reach of channel is an important hydrologic variable because it is related to the ability of a stream to erode its bed and transport sediment. Total or available stream power is proportional to the slope of the water surface and discharge. The gradient of the water surface generally correlates well with the channel slope, and there is also a good correlation between total channel length upstream and bankful discharge (the discharge necessary to fill a channel) which is thought to be important in forming and maintaining rivers.

The *SL* index is very sensitive to changes in channel slope, and this sensitivity allows the evaluation of relationships among possible tectonic activity, rock resistance, and topography. The sensitivity of the *SL* index to rock resistance is illustrated on Figure 4.8, which shows the longitudinal profile of the Potomac River upstream from Washington, D.C. *SL* index values are relatively low in the Valley and Ridge province and in the Appalachian Valley, where the rock types are shale, siltstone, some sandstone, and carbonate rocks. The index increases dramatically where the river crosses the relatively hard rocks of the Blue Ridge, then decreases on relatively soft rocks of the Triassic basin and the Piedmont. Finally, the index increases dramatically again at the resistant rocks

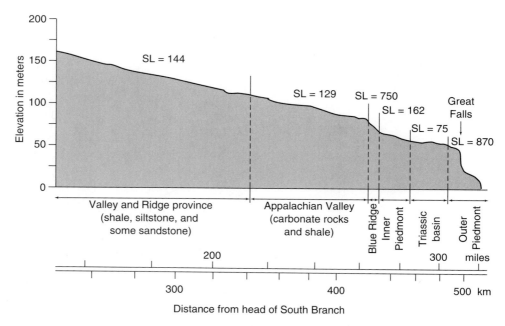

FIGURE 4.8
Stream length-gradient indices *(SL)* for the Potomac River upstream from Washington, D.C.
(After Hack, 1973 [5])

of the Great Falls in the lower reaches of the river [5]. Studies in the Appalachian Mountains of the eastern United States suggest that a good correlation exists between rock resistance and stream length–gradient index. In other words, the form of the land is well adjusted to rock resistance.

In landscape evolution, the adjustment of stream profiles to rock resistance is assumed to occur fairly quickly. Therefore, the *SL* index is used to identify recent tectonic activity by looking for anomalously high index values on a particular rock type. An area of high *SL* indices on soft rock may indicate recent tectonic activity. Anomalously low values of the index may also represent tectonic activity. For example, along linear valleys produced by strike-slip faulting, low indices are expected because the rocks in the valleys are often crushed by fault movement, and the streams flowing through those valleys should have a lesser slope.

A map of the *SL* indices is produced by the following steps:

1. Obtain a topographic map of the area of interest. The method is particularly useful for large areas, so 1:250,000 scale maps often are useful. Using a transparent overlay, outline the major streams and rivers. Extend streams to where contour lines no longer "V" in the upstream direction.

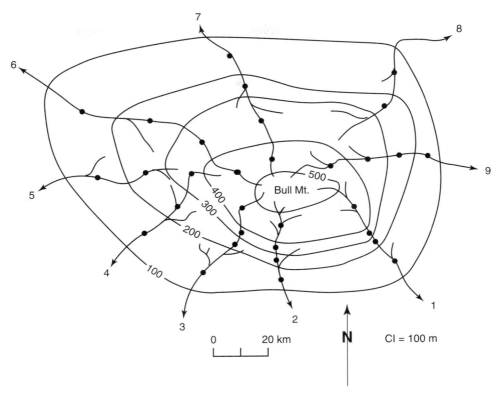

FIGURE 4.9
Hypothetical mountain showing construction of map of level to which streams have
eroded. Dots are points where stream length-gradient index *(SL)* should be calculated.

2. Select a convenient contour interval (say 100 m for a map scale of
    1:250,000). Where these contours cross the mapped streams, mark the
    locations on the overlay.
3. Connect points of equal elevation between the streams. This will create
    a map of the level to which the streams have eroded, often called a
    subenvelope map. Figure 4.9 shows a hypothetical map of the level
    to which streams have eroded Bull Mountain.
4. Along each stream, measure the distance, ΔL, between successive con-
    tours along the stream as well as the total upstream stream length. The
    calculations are simple because the contour interval is constant, and thus
    ΔH is constant. Calculate the *SL* index (Equation 4.4) for each small-
    stream segment, and mark that value at the midpoint between contours
    on the subenvelope map. These locations are noted as dots on Figure 4.9.
    The values of the index farthest upstream, near the drainage divide, may

be spurious, so it may be necessary to start calculating the index a standard distance downstream from the divide.

5. Finally, construct the *SL*-index map by contouring the *SL* values calculated. The subenvelope map for Bull Mountain (Figure 4.9) suggests that a belt of high indices lies between the contour intervals of 200 m and 400 m for streams 1, 2, and 3. This zone may reflect a resistant rock unit, or perhaps an active fault zone along the southern flank of the range. Field work should confirm or negate the interpretation. As discussed below, *SL* index has proven useful as a reconnaissance tool in active tectonics work in several studies [13, 14, 15].

Stream Length–Gradient Indices in the San Gabriel Mountains of Southern California

SL indices can be calculated for areas of several thousand square kilometers or more, utilizing small-scale topographic maps (1:50,000 to 1:250,000). For example, Figure 4.10 shows *SL* indices for the San Gabriel Mountains of Southern

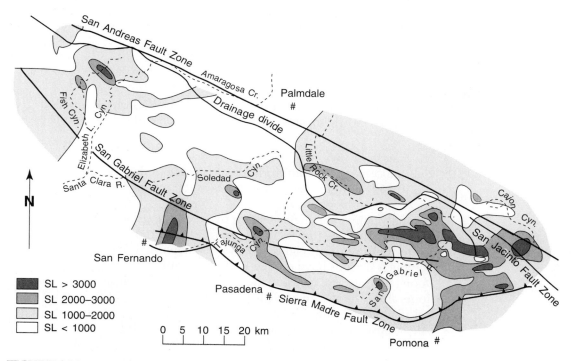

FIGURE 4.10
Stream length–gradient indices *(SL)*, San Gabriel Mountains, Southern California.

California. The mountain range is part of the Transverse Ranges, which have been uplifted from below sea level to several thousand meters elevation in the past few million years. The range is bounded on the north and northeast by the San Andreas fault zone and on the southwest and south by a system of active reverse faults including the San Fernando fault and the Sierra Madre fault zone. The San Gabriel fault zone, also located near the southern and southwestern boundary, is a right-lateral strike-slip fault that is thought to have been more active in the past than at present.

Anomalously high values of the *SL* index occur along the southern and eastern front of the San Gabriel Mountains. Although it has been known for some years that uplift rates are relatively high there, the *SL*-index map of the San Gabriel Mountains demonstrates the utility of the index for identifying zones of tectonic activity. Of particular interest is the zone of high indices near San Fernando. This is the site of the 1971 ($M = 6.6$) San Fernando earthquake that caused widespread damage in the Los Angeles area. Thus, if nothing were known about the San Gabriel Mountains, but good topographic maps and other elevation control were available, the *SL* index would have indicated that the southern and southeastern flanks of the range should be studied in more detail. There the active reverse faults that generated the 1971 earthquake would be found.

Other interesting aspects of the *SL*-index map for the San Gabriel Mountains are the belts of relatively low indices along the San Andreas and San Gabriel fault zones. These index values are most likely related to soft rocks produced by crushing caused by long-term fault movement. Low indices are also found in a belt of rocks with low resistance between Elizabeth Canyon and Soledad Canyon. In the central part of the ranges (at the southeastern end of the map, Figure 4.10), resistant metamorphic rocks are present and high indices are found.

Stream Length–Gradient Indices at the Mendocino Triple Junction, Northern California

A study of tectonic geomorphology at the Mendocino Triple Junction (Figure 4.11) sought to identify zones of active tectonics [15]. This triple junction is the location where the North American, Pacific, and Gorda plates come together at one point. The research evaluated the response of coastal streams to uplift related to the triple junction (also see discussion in Chapter 6). The position of the triple junction is not fixed in time; it is migrating northward at approximately 56 mm/yr—the same rate as the relative movement between the Pacific and North American plates along the San Andreas fault. North of the triple junction, the tectonics are dominated by northeast-directed compression, producing northwest-trending folds and thrust faults [16]. South of the triple junction, the tectonics are related to the right-lateral strike-slip motion of the San Andreas fault system, with localized extension and compression. At the triple junction itself, high rates of uplift have been observed.

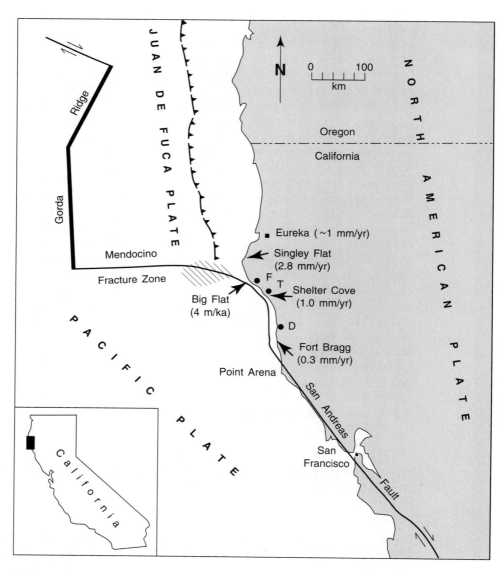

Triple junction

Loction where stream length gradient
data is shown on Figure 4.14

FIGURE 4.11
Tectonic framework of the Mendocino triple junction area (F is Fourmile Creek, T is
Telegraph Creek, and D is DeHaven Creek).
(After Merritts and Vincent, 1989 [15])

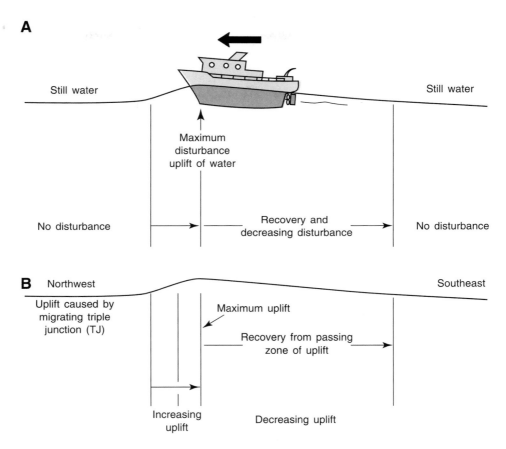

FIGURE 4.12
Idealized diagram to illustrate uplift process related to migrating Mendocino Triple Junction. (A) The moving ship represents the migrating junction disturbing the water surface. (B) The migrating triple junction produces a similar pattern of uplift.

As the triple junction migrates northward, it acts similarly to a ship plowing through calm lake water (Figure 4.12). In front of the ship, uplift of the water surface increases to a maximum near the ship's bow. In the wake of the ship, uplift decreases back to the calm water level. Similarly, uplift rates due to the passage of the triple junction increase from about 1 mm/yr near Eureka, to 2.8 mm/yr at Singley Flat, to a maximum of about 4 mm/yr at Big Flat, and then uplift rates decline gradually to the south. At Fort Bragg, rates are approximately 0.3 mm/yr (Figure 4.11) [15]. The rates of uplift along the coast were established by careful measurement and dating of uplifted marine platforms.

Both climate and rock resistance are relatively uniform along the coast of the study area (Figure 4.11). Detailed study of the coastal streams suggests that

FIGURE 4.13
Relationship between channel gradient of first-order streams and uplift rate near the Mendocino Triple Junction.
(After Merritts and Vincent, 1989 [15])

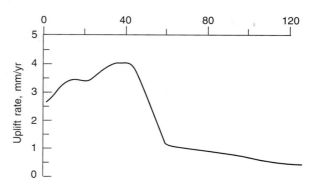

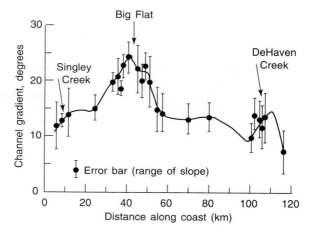

the first-order streams (those farthest headward in the drainage basin) are most sensitive to tectonics and thus are the best indicators of areas with high rates of uplift. Figure 4.13 shows the uplift rate and channel gradients of the first-order stream channels. In contrast, the channel gradients of higher order stream channels (not illustrated) do not correlate well with uplift rates [15]. Evidently larger streams have sufficient stream power to overwhelm the effects of tectonics. For a more detailed explanation of the relationship between stream power and tectonics that flow over bedrock, see Chapter 6.

SL indices were also evaluated for the Mendocino Triple Junction area. The index allowed discrimination between streams characterized by high to intermediate uplift rates and streams characterized by low uplift rates. Figure 4.14 shows the profiles of three streams (approximate locations shown on Figure 4.11) along with average *SL* indices for stream reaches. DeHaven Creek is an area with a low uplift rate, less than 1 mm/yr, and *SL* indices are also relatively low, except

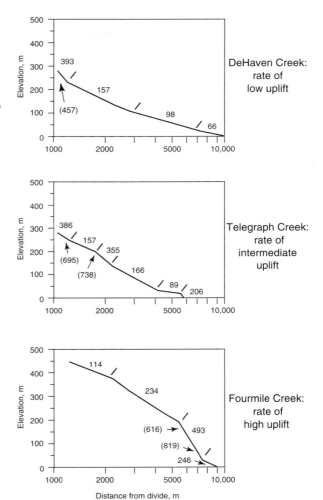

FIGURE 4.14
Stream length–gradient indices *(SL)* for streams in areas of contrasting rates of uplift. Locations shown on Figure 4.11.
(After Merritts and Vincent, 1989 [15])

in the headward part of the stream profile. Telegraph Creek is an area of intermediate rate of uplift, and the *SL* indices are of intermediate value along the entire profile. Finally, Fourmile Creek is in an area of rapid uplift and has relatively high *SL* indices along its entire length, as well as a convex profile which is characteristic of streams undergoing rapid uplift [15].

A conclusion of this study is that the *SL* index is clearly able to distinguish between low, intermediate, and high rates of uplift. In particular, first-order stream channels are most sensitive to recent tectonic activity. This study confirmed the usefulness of the *SL* index as a reconnaissance tool in categorizing relative magnitudes of uplift in an area.

MOUNTAIN-FRONT SINUOSITY

Mountain-front sinuosity [6, 7] is defined as:

$$S_{mf} = L_{mf} / L_s \qquad (4.5)$$

where S_{mf} is the mountain-front sinuosity; L_{mf} is the length of the mountain front along the foot of the mountain, at the pronounced break in slope; and L_s is the straight-line length of the mountain front (Figure 4.15). Mountain-front sinuosity is an index that reflects the balance between erosional forces that tend to cut embayments into a mountain front and tectonic forces that tend to produce a straight mountain front coincident with an active range-bounding fault. Those mountain fronts associated with active tectonics and uplift are relatively straight, with low values of S_{mf}. If the rate of uplift is reduced or ceases, then erosional processes will carve a more irregular mountain front, and S_{mf} will increase.

In practice, the values of S_{mf} may be calculated easily from topographic maps or aerial photographs. However, values of S_{mf} depend on image scale [17], and small-scale topographic maps (1:250,000) produce only a rough estimate of mountain-front sinuosity. Aerial photographs and larger scale maps, with resolution greater than the irregularity of the mountain front, are more useful when calculating S_{mf}.

FIGURE 4.15
Idealized diagram showing how mountain front sinuosity (S_{mf}) is calculated.

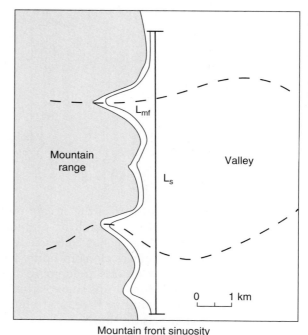

Mountain front sinuosity

$$= S_{mf} \frac{L_{mf}}{L_s} = \frac{10.5 \text{ km}}{8.5 \text{ km}} = 1.2$$

Mountain-Front Sinuosity Near the Garlock Fault, California

One of the first studies that used S_{mf} evaluated relative tectonic activity north and south of the Garlock fault in California [17]. In Figure 4.16, the dashed lines represent the straight-line lengths of the mountain fronts that were evaluated. The solid lines are the outlines of the actual mountain fronts,

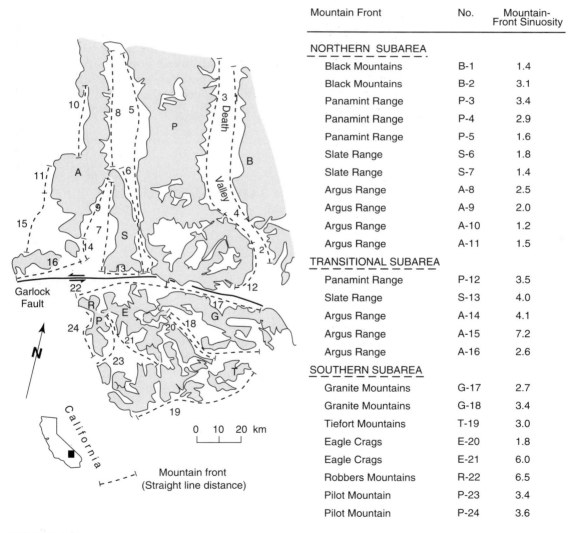

Mountain Front	No.	Mountain-Front Sinuosity
NORTHERN SUBAREA		
Black Mountains	B-1	1.4
Black Mountains	B-2	3.1
Panamint Range	P-3	3.4
Panamint Range	P-4	2.9
Panamint Range	P-5	1.6
Slate Range	S-6	1.8
Slate Range	S-7	1.4
Argus Range	A-8	2.5
Argus Range	A-9	2.0
Argus Range	A-10	1.2
Argus Range	A-11	1.5
TRANSITIONAL SUBAREA		
Panamint Range	P-12	3.5
Slate Range	S-13	4.0
Argus Range	A-14	4.1
Argus Range	A-15	7.2
Argus Range	A-16	2.6
SOUTHERN SUBAREA		
Granite Mountains	G-17	2.7
Granite Mountains	G-18	3.4
Tiefort Mountains	T-19	3.0
Eagle Crags	E-20	1.8
Eagle Crags	E-21	6.0
Robbers Mountains	R-22	6.5
Pilot Mountain	P-23	3.4
Pilot Mountain	P-24	3.6

FIGURE 4.16

Mountain-front sinuosity (S_{mf}) north and south of the Garlock fault.
(After Bull and McFadden, 1977 [17])

reflecting their sinuosity. The values of sinuosity seem to define three groups of activity [17]:

- the subarea north of the Garlock fault, with relatively low S_{mf}
- a transitional subarea in the central part of the map, north of and adjacent to the Garlock fault, with higher values of S_{mf}
- a subarea south of the Garlock fault, with relatively high values of S_{mf}

This study concluded that the most active mountain fronts—those associated with active, range-bounding faults—generally have an S_{mf} between 1.0 and 1.6. Mountain fronts with lesser activity, but still reflecting active tectonics, have sinuosities between approximately 1.4 and 3. Inactive mountain fronts have sinuosities from about 1.8 to greater than 5. In general, sinuosity values greater than 3 are associated with fronts that are so eroded and embayed that the topographic range fronts, at one time coincident with the active geologic structures, are now 1 km or more away [17].

Like the stream length–gradient index, mountain-front sinuosity is a potentially valuable reconnaissance tool used to identify areas of tectonic activity.

RATIO OF VALLEY-FLOOR WIDTH TO VALLEY HEIGHT

The ratio of valley floor width to valley height (V_f) may be expressed as:

$$V_f = 2V_{fw} / [(E_{ld} - E_{sc}) + (E_{rd} - E_{sc})] \tag{4.6}$$

where V_f is the valley-floor width-to-height ratio, V_{fw} is the width of the valley floor, E_{ld} and E_{rd} are elevations of the left and right valley divides, respectively, and E_{sc} is the elevation of the valley floor [6, 7] (Figure 4.17). When calculating V_f, these parameters are measured at a set distance from the mountain front for every valley studied. This index differentiates between broad-floored canyons, with relatively high values of V_f, and V-shaped valleys with relatively low values. High values of V_f are associated with low uplift rates, so that streams cut broad valley floors. Low values of V_f reflect deep valleys with streams that are actively incising, commonly associated with uplift.

The study that evaluated the mountain-front sinuosity north and south of the Garlock fault also calculated values of the ratio of valley-floor width to valley height [17]. The V_f values in that area ranged from 0.05 to 47.0. Lower values were associated with valleys north of the Garlock fault, where tectonic activity is assumed to be more vigorous.

RELIC MOUNTAIN FRONTS

Sinuosity and the ratio of valley-floor width to height establish that some mountain fronts are more active than others. Older mountain fronts, located *within*

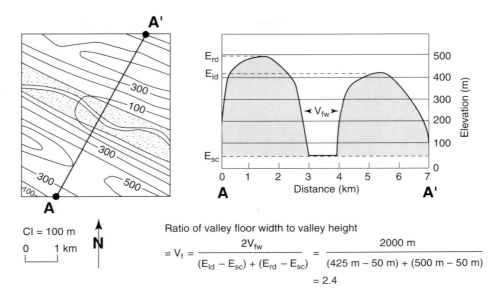

FIGURE 4.17
Idealized diagram illustrating how the ratio of valley-floor width to valley height (V_f) is calculated. Note: Left and right is determined by looking downstream.

ranges, commonly are no longer active. Hypothetically, mountain fronts formed early in a range's development are active for a period of time, then deformation migrates toward the edges of the range, and new mountain fronts form. In the western Transverse Ranges of California, we see a central highlands with a number of inactive mountain fronts is flanked by fold-and-thrust belts that are presently active (Figure 4.18). The interior, relic mountain fronts had the same basic morphology as the active, outer fronts—streams that emerged from the fronts commonly fed a series of alluvial fans on the margins of the ranges. When tectonic activity shifted, the old alluvial fans were sometimes consumed by the active mountain-building process as new fronts developed. Thus, in a sense, as the mountain ranges grow outward, they consume their own alluvial fans.

CLASSIFICATION OF RELATIVE TECTONIC ACTIVITY

Each of the indices discussed provides a relative classification of tectonic activity useful in reconnaissance studies. When more than one index is applied to a particular region, the results are more meaningful than those of any single analysis. This concept was tested by evaluating mountain fronts on the southern flank of the central Ventura Basin in California (Figure 4.19). The basin has a complex geologic history characterized by extension and rotation during the Miocene and

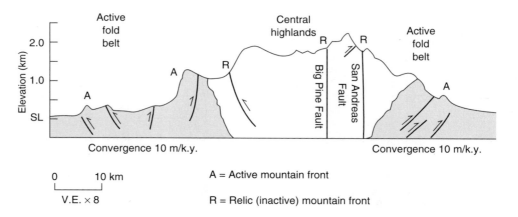

FIGURE 4.18

Active and relic mountain fronts, Western Transverse Ranges. The San Andreas and Big Pine faults are right-lateral and left-lateral strike-slip faults, respectively.

strong compression from the Pliocene to the present [18]. Eight mountain fronts in the area were evaluated in terms of S_{mf} and V_f. In addition, SL indices were calculated for the same area. All of the mountain fronts are bounded by active folds or faults. Mountain fronts associated with the San Cayetano fault (fronts 4–8 on Figure 4.19; Figure 4.20) generally have high SL indices, whereas the fronts associated with the Ventura Avenue anticline and the buried Ventura fault (fronts 1–3) have relatively low indices. S_{mf} of the young mountain front at the San Cayetano fault is 1.14, and the V_f is 0.47, reflecting relatively high active tectonics. The reason for the low SL indices along front 1 (Figure 4.19) is that the rocks are weak, and the gradients of streams that cross these rocks do not reflect the tectonic activity in that area. However, the other indices of relative tectonic activity do indicate active deformation; the front has a very low mountain-front sinuosity and low valley-floor width-to-height ratios. Table 4.1 summarizes selected geomorphic parameters for the eight mountain fronts.

When all the information from the mountain fronts and SL indices is combined, it is possible to produce a **relative tectonic activity class designation** [6, 7]. Mountain fronts that are suggestive of the highest tectonic activity are designated as class 1 fronts. These fronts typically have low values of S_{mf}, low V_f, and high SL indices. In the central Ventura Basin, these fronts are usually associated with an uplift rate of greater than 1 mm/yr. Class 2 mountain fronts are associated with less tectonic activity, reflected in higher S_{mf} and V_f values and lower SL values. Class 3 mountain fronts are still associated with active tectonics, but geomorphic indices suggest less activity than for class 2 fronts. An example of a class 2 front in the central Ventura Basin is mountain front 6, which has an S_{mf} of 2.7 and a V_f of approximately 1.9. Mountain fronts characterized by minimal tectonic activity or those that are now inactive may be classified as class 4 or class 5 [6, 7].

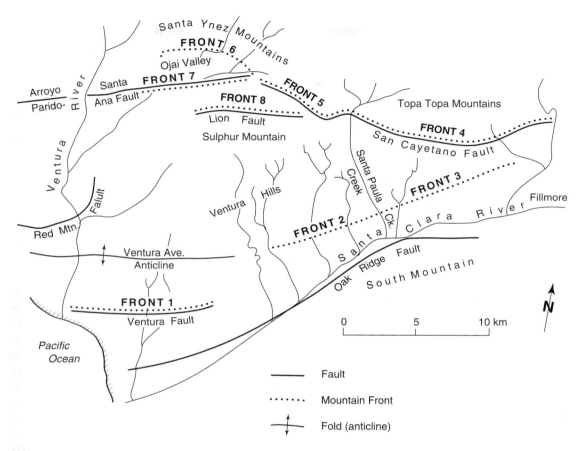

FIGURE 4.19
Mountain fronts and major structures used for the tectonic geomorphic analysis in the central Ventura Basin. (Values of S_{mf}, V_f and tectonic activity class for these mountain fronts are listed on Table 4.1.)
(After Rockwell et al., 1985. In Morisawa and Hack (eds.), *Tectonic Geomorphology.* Allen & Unwin: Boston)

Attempting to classify mountain fronts in terms of relative tectonic activity is a fairly new endeavor [6]. The classification and boundaries between classes stated above are arbitrary and only indicate relative differences. No attempts have been made to put specific bounds on values of geomorphic indices, because these indices reflect local conditions of rock type, structure, and climate. Their real usefulness is for differentiating mountain fronts as very active, moderately active, or inactive. This information is valuable in areas where detailed field studies have not yet been conducted.

TABLE 4.1

Locations of mountain fronts of the Western Transverse Ranges shown in Figure 4.19.

Front Number	Location of Tectonic Mountain Front	Type of Front	Orientation of Front	Front Length (km)	Maximum Relief (m)	Mountain-Front Sinuosity S_{mf}	Valley-Floor Width to Valley Height V_f	Tectonic Activity Class	Uplift Rate
1	Ventura River to Harmon Cyn.	Bounding folded (VAA)	E–W	6.45	438	1.09	0.7	1	4 mm/yr
2	Aliso Cyn. to Fagon Cyn.	Bounding folded	N60E	8.88	670	1.57	1.8	2	unknown
3	Orcutt Cyn. to Snow Cyn.	Bounding folded	N60E	10.1	1280	1.83	1.91	2	unknown
4	S.P. Creek to Sespe Creek	Internal faulted (SCF)	E–W	6.45	915	1.14	0.43	1	2–8 mm/yr
5	Sisar to Santa Paula Creek	Internal/bounding faulted (SCF)	N60W	5.65	1639	1.14	0.47	1	0.5–1.5 mm/yr
6	Wilsie to West Gridley Canyon	Bounding folded	Curving front NE–W	5.25	1363	2.72	1.89	2–3	unknown
7	Wilsie to San Antonio Creek	Bounding (APF) faulted	N75E	5.85	305	1.01	<0.73	1	0.4 mm/yr
8	South side Upper Ojai Valley	Bounding faulted (Lion F.)	N80E to EW	4.23	451	1.34	0.80	1	unknown

(After Rockwell et al., 1985. In M. Morisawa and J.T. Hack (eds.), *Tectonic Geomorphology.* Allen & Unwin: Boston)

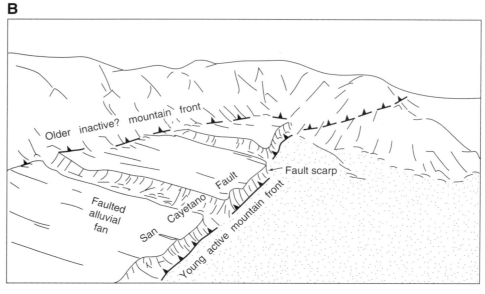

Thrust fault

FIGURE 4.20
(A) Oblique aerial photograph of truncated alluvial fan at the San Cayetano fault along mountain front 5 (Figure 4.19). (B) Sketch map illustrating points of interest on (A). The nearly-straight fault scarp at this location is approximately 60 m high. Upstream of this location (in upper left part of the photo) is an older mountain front that may no longer be active. (Photograph by E.A. Keller and T. Rockwell)

SUMMARY

Quantitative measurements have allowed geomorphologists to objectively compare landforms and calculate geomorphic indices that are useful for identifying a particular characteristic of the area; for example, its level of tectonic activity.

The hypsometric curve and hypsometric integral are related to the degree of dissection of a landscape. Hypsometric analysis is a useful tool for differentiating tectonically-active from tectonically-inactive regions. Drainage-basin asymmetry is defined in terms of the Asymmetry Factor as well as the Transverse Topographic Symmetry Factor. Both of these are valuable in rapid evaluations of drainage basins to determine if tectonic tilt may have occurred. The stream length–gradient index (SL) is a useful tool for studying tectonic geomorphology; high values commonly are found where streams cross resistant rocks or where streams cross active structures. Values of the stream length–gradient index may be calculated easily from topographic maps to provide information where more detailed studies would be fruitful. Mountain-front sinuosity (S_{mf}) is an index that reflects the balance between erosional forces and tectonic forces at a mountain front. In general, active mountain fronts have relatively low values of sinuosity. The ratio of valley-floor width to valley height (V_f) differentiates between broad-floored canyons and V-shaped valleys. Low values of V_f are associated with active tectonics.

When several indices of relative tectonic activity are evaluated for a particular region, it is possible to develop a system of relative tectonic-activity classes. Commonly, it is useful to classify areas as being very active, moderately active, or inactive. Such basic classification is useful in delineating areas where more detailed field studies will identify active structures and calculate rates of active tectonic processes.

REFERENCES CITED

1. Keller, E.A., 1986. Investigations of active tectonics: use of surgical earth processes. In *Panel on Active Tectonics*. National Academy Press: Washington, D.C.

2. Strahler, A.N., 1952. Hypsometric (area-altitude) analysis of erosional topography. *Geological Society of America Bulletin*, 63: 1117–1142.

3. Hare, P.W., and T.W. Gardner, 1985. Geomorphic indicators of vertical neotectonism along converging plate margins, Nicoya Peninsula, Costa Rica. In M. Morisawa and J.T. Hack (eds.), *Tectonic Geomorphology: Proceedings of the 15th Annual Binghamton*

Geomorphology Symposium, September 1984. Allen & Unwin: Boston.

4. Cox, R.T., 1994. Analysis of drainage basin symmetry as a rapid technique to identify areas of possible Quaternary tilt-block tectonics: an example from the Mississippi Embayment. *Geological Society of America Bulletin*, 106: 571–581.

5. Hack, J.T., 1973. Stream-profile analysis and stream-gradient index. *U.S. Geological Survey Journal of Research*, 1: 421–429.

6. Bull, W.B., 1977. Tectonic geomorphology of the Mojave Desert. *U.S. Geological Survey Contract Report* 14-08-001-G-394. Office of

Earthquakes, Volcanoes, and Engineering: Menlo Park, CA.

7. Bull, W.B., 1978. Geomorphic tectonic classes of the south front of the San Gabriel Mountains, California. *U.S. Geological Survey Contract Report* 14-08-001-G-394. Office of Earthquakes, Volcanoes, and Engineering: Menlo Park, CA.

8. Pike, R.J., and S.E. Wilson, 1971. Elevation-relief ratio, hypsometric integral and geomorphic area-altitude analysis. *Geological Society of America Bulletin,* 62: 1079–1084.

9. Mayer, L., 1990. *Introduction to Quantitative Geomorphology.* Prentice Hall: Englewood Cliffs, NJ.

10. Harlin, J.M., 1978. Statistical moments of the hypsometric curve and its density function. *Mathematical Geology,* 10: 59–72.

11. Gardner, T.W., K.C. Sasowski, and R.L. Day, 1990. Automated extraction of geomorphometric properties from digital elevation data. *Zeitshrift fur Geomorphologie Supplement,* 80: 57–68.

12. Gardner, T.W., W. Back, T.F. Bullard, P.W. Hare, R.H. Kesel, D.R. Lowe, C.M. Menges, S.C. Mora, F.J. Pazzaglia, I.D. Sasowski, J.W. Troester, and S.G. Wells, 1987. Central America and the Caribbean. In W.L. Graf (ed.), *Geomorphic Systems of North America, Centennial Special Volume 2.* Geological Society of America: Boulder, CO.

13. Keller, E.A., 1977. Adjustments of drainage to bedrock in regions of contrasting tectonic framework. *Geological Society of America Abstracts with Programs,* 9: 1046.

14. Zhao, X., 1990. Tectonic geomorphology and soil chronology of the Frazier Mountain area, Western Transverse Ranges, California. Ph.D. dissertation. University of California, Santa Barbara, California.

15. Merrits, D., and K.R. Vincent, 1989. Geomorphic response of coastal streams to low, intermediate, and high rates of uplift, Mendocino Triple Junction region, Northern California. *Geological Society of America Bulletin,* 110: 1373–1388.

16. Clark, S., and G. Carver, 1992. Late Holocene tectonics and paleoseismicity, Southern Cascadia subduction zone. *Science,* 255: 188–192.

17. Bull, W.B., and L.D. McFadden, 1977. Tectonic geomorphology north and south of the Garlock fault, California. In D.O. Doehring (ed.), *Geomorphology in Arid Regions. Proceedings of the Eighth Annual Geomorphology Symposium.* State University of New York at Binghamton, Binghamton, NY.

18. Keller, E.A., and T.K. Rockwell, 1984. Tectonic geomorphology, Quaternary chronology and paleoseismology. In J.E. Costa and P.J. Fleisher (eds.), *Developments and Applications of Geomorphology.* Springer-Verlag: Berlin.

5

Active Tectonics and Rivers

The face of [the river], in time, became a wonderful book . . . which told its mind to me without reserve, delivering its most cherished secrets as clearly as if it uttered them with a voice. . . . In truth, the passenger who could not read this book saw nothing but all manner of pretty pictures in it, painted by the sun and shaded by the clouds, whereas to the trained eye these were not pictures at all, but the grimmest and most dead-earnest of reading matter.

Mark Twain
Life on the Mississippi

FIGURE 5.1
The Norman River in the North Queensland Gulf Region of Australia.
(Photo by Bill Bachman/Photo Researchers, Inc.)

INTRODUCTION

Of all the processes acting on the surface of the Earth, rivers are neither the most powerful (compared, for example, to glaciers or volcanoes) nor the most widespread (compared, for example, to gravity-driven slope processes). Yet rivers have a special place in the human world and in the human psyche—in the locations of our cities, as the avenues of commerce, as the life-blood of agriculture, and in our poetry and literature. Perhaps it is for this reason that, of all the geomorphic systems, so much study and research has gone into river systems. At the same time, rivers also have all the makings of a scientific obsession—they are tremendously varied and complex, but also systematic—as if just one more crucial insight would explain all of the variety and complexity.

The study of river systems is known as **fluvial geomorphology** (from the Latin *fluvius,* meaning "river"). The forms of rivers or streams and the processes occurring in those systems are described by a large number of parameters: channel width and depth, dissolved sediment load, suspended load, bed load, channel slope and sinuosity, flow velocity, channel roughness, and many others. The delicate balance between all of these parameters in a river system means that rivers are very sensitive to any kind of *change.* Climatic changes that have repeatedly swept over the Earth during the Quaternary Period have had profound effects on most geomorphic systems, including rivers [1]. Changes in global sea level over the same interval (see Chapter 6) have caused large-magnitude cycles of **aggradation** (accumulation of sediment in river valleys) and **degradation** (removal of material from river valleys). Last but not least, the single most effective agent of geomorphic change has been humans, who in their brief tenure upon the Earth have nearly doubled the sediment supply to the world's rivers [2] and otherwise monkeyed with fluvial systems worldwide. In addition, with the growth of the science of tectonic geomorphology has come the growing realization that active tectonic process also can influence river form and process [3, 4, 5].

Coseismic Modification of River Systems

As students of both tectonic and fluvial geomorphology, we are interested in the changes wrought by earthquakes on river systems. Among such changes, some occur *during* earthquakes as direct results of ground shaking or coseismic deformation. Many of these *coseismic* modifications of rivers present an immediate threat to humans in the vicinity of the rivers, a threat that often exceeds the threat from the shaking itself. Earthquakes may cause sudden shifts in river flow, posing a flood hazard, or they may even permanently alter the course of the river. The Owens River of eastern California brings rainfall and snow melt from the peaks of the Sierra Nevada down to the desert of the Owens Valley. In 1872, a $M = 8$ earthquake struck the Owens Valley and, according to eyewitness reports, "the disturbance of the water in the river . . . was so severe that fish were thrown out upon the bank; and men stopping there, who were engaged in building a boat, did not hesitate to capture them, and served them up for breakfast" [6]. In 1872, the Owens River system included Owens Lake, a small inland lake which was later drained to supply water to the Los Angeles region. During the great earthquake, it was reported that "the water had receded from the shore, and that it stood in a perpendicular wall. . . . The wave, however, returned to shore in the course of two or three minutes, breaking and flowing some two hundred feet beyond the former edge of the shore" [6].

A great earthquake also struck India in 1819. It caused a 6-m uplift that entirely blocked an eastern branch of the Indus River, one of the major rivers of the world. It is reported that for years after the earthquake, the Indus River was "unsettled in its bed." Finally, in 1826, the river underwent sudden **avulsion** (it jumped from one channel to another) and overran the barrier which had been uplifted by the earthquake, regaining its most direct route to the ocean [7].

Gradual Change in River Systems

Of the many effects of active tectonics discussed in this chapter, the majority are neither so sudden nor so dramatic as the coseismic events described above. More typically, tectonic deformation of the Earth's surface takes place slowly over thousands of years or longer. Recall from Chapter 3 that, although deformation is imperceptible to human eyes, it often can be measured by the most sensitive scientific instruments. In their own way, rivers are almost as sensitive to tectonic deformation as are geodetic survey instruments or satellite locating systems. Geomorphologists have used the characteristics of river systems to confirm or refute geodetic measurements that suggest areas of active deformation [3, 8]. After geodetic instruments, which can detect gradual deformation over a few years to a few decades, river systems are the next most sensitive tool, capable of adjusting to deformation over periods of decades to centuries.

TECTONIC MODIFICATION OF ALLUVIAL RIVERS

One of the most fundamental subdivisions of river systems is between (1) those that flow over bedrock channels, and (2) those that flow on a bed of **alluvium** (river sediments). Alluvial rivers are those that "flow between banks and on a bed composed of sediment that is transported by the river" [9]. The form of a given river depends upon the balance between driving forces (gravity, amount of precipitation in the drainage basin) and resisting forces (substrate strength, friction). In alluvial rivers, resisting forces are greater than driving forces, and the river cannot transport all of the available sediment—the result is that the river flows in a bed of its own detritus. In bedrock rivers, driving forces are greater than resisting forces, and all sediment supplied can be transported away, with the result that the river flows over a channel of exposed bedrock. In general, bedrock rivers are associated with smaller drainage basins, higher relief, and stronger bedrock.

In general, alluvial rivers obey a stricter set of rules than do bedrock rivers because of the balance between driving forces and resisting forces. A river that maintains itself in such a state of **dynamic equilibrium** (see Chapter 2) is called a **graded river.** The concept of graded rivers was developed by Mackin [10] and refined by Leopold and Maddock [11]:

> *A graded river is one in which, over a period of years, slope and channel characteristics are delicately adjusted to provide, with available discharge, just the velocity required for the transportation of the load supplied from the drainage basin. The graded stream is a system in equilibrium; its diagnostic characteristic is that any change in any of the controlling factors will cause a displacement of the equilibrium in a direction that will tend to absorb the effect of the change.*

The key to understanding river grade is the last sentence in the definition above—any change in one variable in a graded river system will cause change in other variables in the system in order to reestablish equilibrium.

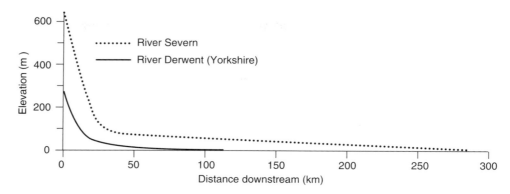

FIGURE 5.2
Equilibrium longitudinal profiles of two rivers in Great Britain. The concave-upward shapes of these profiles are characteristic of most graded rivers.
(After Richards, 1982. *Rivers: Form and Process in Alluvial Channels.* Methuen & Co.: New York)

Longitudinal Profiles

Graded rivers have a characteristic longitudinal profile from their headwaters to their mouths (Figure 5.2), because the different tributaries gradually coalesce, and the volume of flow in the trunk channel increases downstream. If all of the flow in a river were added at the head of just one tributary, then the graded profile of the river would be an enormous waterfall at the headwaters and a uniform, gentle slope from there to the ocean. Longitudinal profiles are also powerful tools for detecting subtle perturbations along a river's course. The **stream length–gradient index** outlined in Chapter 4 is a practical method for measuring such perturbations.

For example, the 1886 Charleston, South Carolina, earthquake ($M_b = 6.6–6.9$) was the largest on the eastern coast of the United States in history, but the exact location of the fault that caused the earthquake is unknown because it did not rupture the surface. However, longitudinal profiles of rivers across the epicentral area are anomalously shaped; they are convex upward across a northeast-trending zone above the buried Woodstock fault [12, 13]. The implication is that rupture on the Woodstock fault was responsible for the 1886 earthquake, but rupture did not penetrate the thick sedimentary cover of the South Carolina coastal plain. Rupture at depth resulted in a belt of muted uplift at the surface [13].

River Pattern

Rivers and streams span a continuous range of forms, a range that we traditionally divide into three classes that depend on valley slope (steepness): (1) straight, (2) meandering, and (3) braided channels (Figure 5.3). Tectonic uplift or tilting

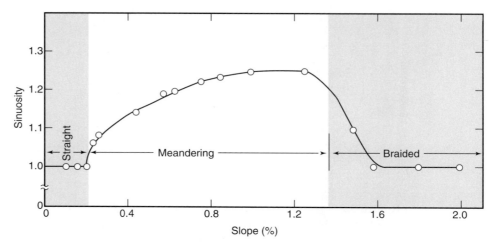

FIGURE 5.3
Relationship between valley slope and sinuosity (channel length/valley length) for a
given discharge. Channels are classified as either straight, meandering, or braided.
(After Schumm and Kahn, 1972. *Geological Society of American Bulletin*, 83: 1755–1770)

of just a few tenths of a percent can change the pattern of a river, such as from
meandering to braided (a 0.1% change in slope is equivalent to differential uplift
of 1 m over a distance of 1 km). The Sefid Rud River in northwest Iran has both
straight and meandering reaches. The largest meandering reach coincides with
the epicentral region of a $M_w = 7.3$ earthquake in 1990 that killed 40,000 people.
It is interpreted that uplift during previous earthquakes and between earthquakes has caused upwarping and higher slopes that have altered the river's
course from straight to meandering [14]. This evidence suggests that upwarping
in the region has been occurring for a long time and probably is caused by the
same process that caused the 1990 earthquake [14].

Sinuosity

Even small amounts of deformation can change the sinuosity of a meandering
river. Sinuosity is measured as illustrated in Figure 5.4. In the conceptual
framework of a graded system, rivers meander in order to maintain a channel
slope in equilibrium with discharge and sediment load. A river meanders
when the straight-line slope of the valley is too steep for equilibrium—the sinuous path of the meanders reduces the slope of the channel. Any tectonic
deformation that changes the slope of a river valley results in a corresponding
change in sinuosity to maintain the equilibrium channel slope. A secondary
effect of this adjustment is that, as a river switches from one sinuosity to
another, the rates of meander migration and floodplain reworking accelerate

FIGURE 5.4
Definition and calculation of
the sinuosity of a river
channel.

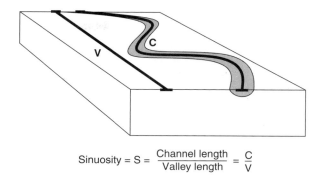

$$\text{Sinuosity} = S = \frac{\text{Channel length}}{\text{Valley length}} = \frac{C}{V}$$

accordingly; this secondary effect itself has proved to be a diagnostic tool in
identifying areas of active tectonics [15].

Repeated geodetic surveys since the late 1800s along the Mississippi
River between Cairo, Illinois, and St. Louis, Missouri, show relative elevation
changes of several tens of centimeters [3]. This amount of deformation is
large for the stable interior of the North American Plate and generally has
been considered to be erroneous. However, the pattern of the Mississippi
reveals a significant correlation between sinuosity and the geodetic results
(Figure 5.5). The river is most sinuous in zones of purported downtilt, and
least sinous in zones of uptilt, just as expected. The river pattern provides
independent evidence that the geodetic measurements are valid and the tectonic deformation is real [3].

BEDROCK-CHANNELED RIVERS

In a bedrock river, driving forces exceed resisting forces, and the channel tra-
verses bare bedrock. One of the defining studies on the effects of recent tectonic
activity in bedrock river systems was described in detail in Chapter 4 (Stream
Length–Gradient Indices at the Mendocino Triple Junction, Northern California) [16]. Small drainage systems along the Pacific coast of northern California
reflect the history of tectonic deformation in that area. The Mendocino Triple
Junction, where the Pacific, North American, and Juan de Fuca plates meet
(see Figure 6.1), is a zone of high uplift rates—4 mm/yr, versus 0.3 mm/yr
away from the triple junction [17]. Through time, northward migration of the
Mendocino Triple Junction has caused a wave of uplift to move along the coast.
As discussed in Chapter 4, the gradient (steepness) of the bedrock channels is
greatest near the region of greatest uplift rate.

The effect of uplift rates, however, is not evenly distributed through each
individual drainage basin. The smallest, most upstream tributaries in each basin
"feel" the effect of uplift the most, and the downstream trunk channels "feel"
the effects the least (Figure 5.6) [16]. The explanation for this is illustrated in

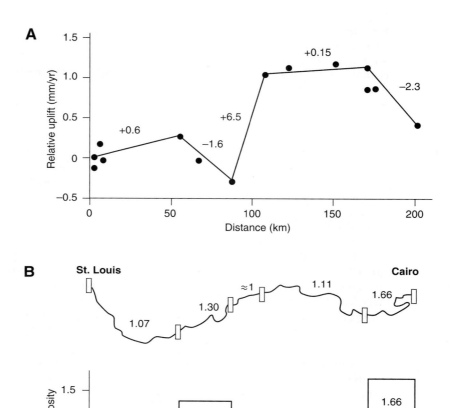

FIGURE 5.5
(A) Relative uplift and subsidence along the Mississippi River between St. Louis, Missouri, and Cairo, Illinois, as determined by geodetic releveling. (B) Average sinuosity of the Mississippi for the same intervals measured in (A) above. Note that where the geodetic measurements indicate uptilt, the sinuosities are relatively low; where there is downtilt, sinuosities are relatively high.
(After Adams, 1980 [3])

Figure 5.7. In a setting undergoing uplift, the rate *(R)* at which a point on a river increases in altitude through time is equal to

$$R = \text{uplift rate} - \text{rate of incision}. \qquad (5.1)$$

Even though entire drainage basins are being uplifted, the largest streams have the most energy for erosion and are best able to maintain equilibrium profiles (Figure 5.7). The largest streams will not increase in altitude through time as much as the small tributaries, nor will their gradients increase as much if the rate of uplift increases.

FIGURE 5.6

Plots of (A) uplift rate and (B) channel gradients of first-, second-, and third-order channels along a transect along the coast of northern California near the Mendocino Triple Junction. Note that the channels are steepest where the uplift rate is greatest, and that the first-order gradients show the most pronounced effect.

(From Merritts and Vincent, 1989 [16])

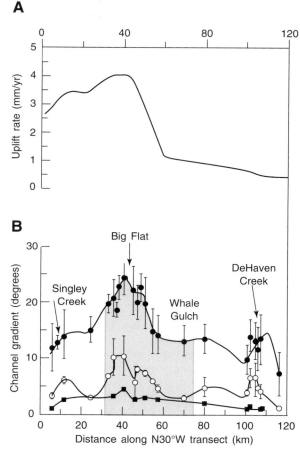

The conclusion that gradient variations in different drainage basins near the Mendocino Triple Junction are caused by active tectonics is valid only if other hypotheses can be excluded. Systematic change in climate could systematically alter fluvial systems, but it is argued that long-term climate change was not systematic in northern California during the period of time in question [16]. Another plausible explanation for the variations in channel gradients along the northern California coast is differences in the local bedrock. In this region, different rock types do crop out and bedrock strength does seem to influence drainage density (the total number of channels in a basin) [18], but the differences do not correspond to the differences in channel gradient [16].

An important contribution from river studies near the Mendocino Triple Junction is a sense of the time required for a geomorphic response to a tectonic stimulus. Because the triple junction and its associated pulse of uplift are moving

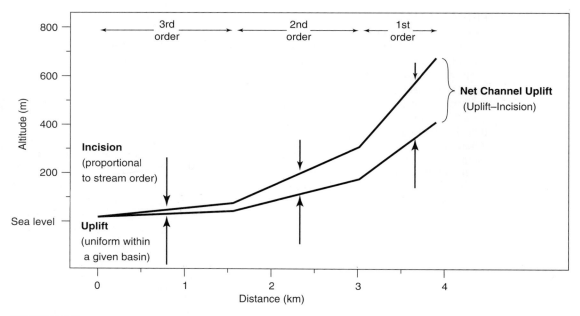

FIGURE 5.7
Illustration of why first-order channels are uplifted most and become steepest as a result
of uplift. The change in elevation of a point on the channel is the net effect of uplift (uniform across the drainage basin) and erosion of the channel (greater farther downstream,
where the channel carries more water).

at a fixed rate—5.6 cm/yr [19]—distance along the coast can be translated into
time. The streams with the steepest gradients are about 7.3 km away from the
zone of the greatest uplift rates—suggesting that the drainage systems require
about 130 k.y. (at 5.6 cm/yr) to re-equilibrate to accelerated uplift. This estimated
response time is specific to rivers in this study area; basins of a different size, in a
different climate, or with a different assemblage of rocks will respond differently.
However, time is a very difficult variable to quantify, and the 130-k.y. lag time
remains one of the firmest estimates available. The issue of the balance between
uplift and degradation is more fully elaborated in Chapter 9.

EFFECTS OF BASE LEVEL

Base level refers to the lowest elevation in a drainage basin, at the foot or outlet
of a channel network. A fluvial system can degrade no lower than its base level
and, if necessary, will aggrade to remain in equilibrium with that level. For continental-scale drainage systems, such as the Mississippi River, base level is the
ocean. Sea level is sometimes referred to as **ultimate base level** because all

rivers that flow to the coast reach that identical lowest level. Some streams do not reach the coast, though; for example, streams that drain into Death Valley. The lowest elevation on those streams is a **local base level**—86 meters below sea level in the case of Death Valley. Local base level also refers to any portion of a river, such as where it flows over a very resistant rock unit, that controls degradation or aggradation upstream.

Base level, both local and ultimate, can be influenced by tectonics and other processes, especially Ice Age sea-level change. Base-level change exerts a major influence on river systems. The primary mechanisms by which base level influences river form are (1) the change in the system's potential energy, and (2) the sediment supply [20]. The response to a base-level rise is aggradation. A river will respond in one of two ways to a base-level fall: (1) by incising, or (2) by changing internal variables, such as sinuosity or channel pattern, with little or no incision [21]. The more rapid the rate of base-level fall, the more likely that the river will incise, confine itself in a narrow channel, and transmit the effects of the base-level fall upstream [21].

DRAINAGE NETWORKS AND DRAINAGE PATTERN

A drainage network formed on a stable, homogeneous landscape will develop a characteristic pattern known as a **dendritic** drainage pattern (Figure 5.8A). In the real world, dendritic patterns may develop in small drainage basins, but variations in bedrock strength and geologic structures tend to result in different patterns at regional scales. Of the different drainage patterns, several indicate distinct bedrock structures. For example, a trellis pattern (Figure 5.8C) can form as a result of strong jointing in the rock. This kind of structural control over the drainage pattern does not imply any active deformation, only heterogeneity in the bedrock. In contrast, some drainage patterns result from active deformation. For example, radial drainage (Figure 5.8D) typically are formed by ongoing upwarp of a small dome, such as caused by salt diapirs (columns of salt that rise through denser surrounding sediment). The first great petroleum reservoirs of the Gulf of Mexico region and the Middle East were found around salt diapirs (oil and gas are trapped against the margins of the salt), because diapirs may be located by surface expressions such as drainage pattern.

Offset Streams

A characteristic feature of a strike-slip fault is that small streams are offset across the fault with a distinct right-lateral or left-lateral sense (Figure 5.9) (also see Landforms of Strike-Slip Faulting in Chapter 2). Offset streams reflect movement on the fault during one or more earthquakes and/or by gradual,

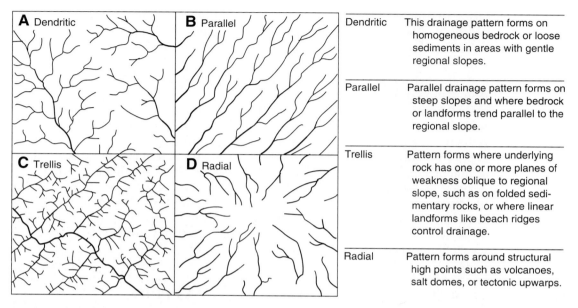

FIGURE 5.8
Classification of the basic drainage patterns.
(After Bloom, A.L., 1991. *Geomorphology: A Systematic Analysis of Late Cenozoic Landforms.* Prentice-Hall: Englewood Cliffs, NJ)

aseismic motion. Where offset channels can be dated, they can provide estimates of the rates of fault slip.

One of the most famous offset streams is Wallace Creek, which is offset nearly 400 m across the San Andreas fault at a site approximately 150 km northwest of the city of Los Angeles. Wallace Creek (Figure 5.9) is an ephemeral stream (a stream which flows only during times of precipitation) that flows at nearly right angles to the San Andreas fault [22]. Figure 5.10 shows a series of sketch maps depicting the recent history of faulting at Wallace Creek, where trenches were excavated to determine the sequence of deposition and erosion. This sequence allowed determination of slip rates for this section of the San Andreas fault. Age control was provided by carbon-14 analysis of organic material in the stream alluvium. Figure 5.10 shows five stages in the development of Wallace Creek [22]:

- Stage 1: deposition of alluvium dated at approximately 19 ka
- Stage 2: channel incision sometime before about 5.9 ka
- Stage 3: offset of Wallace Creek by approximately 250 m between 5.9 ka and 3.9 ka
- Stage 4: the offset stream was abandoned in favor of a more direct route across the fault sometime during the last 3900 yr
- Stage 5: continued offset along the fault produced 380 m of offset on the Stage 2 channel and 130 m of offset on the Stage 4 channel between 3900 ka and present

FIGURE 5.9
Oblique aerial photograph of streams offset across the San Andreas fault at
Wallace Creek.
(From Wallace, R., 1990. *U.S. Geological Survey Professional Paper* 1515)

Reasoning that the total offset of 380 m must be older than or approximately equal
to the oldest materials deposited after the first stream incision (>5900 ka), the
380 m of offset and the 5900 yr interval provide a maximum slip rate of 64 mm/yr.
The 130-m offset occurred in ≤3900 yr, providing a minimum slip rate of 33 mm/yr.
Taken together, these two conclusions suggest that the long-term slip rate at Wallace Creek is between 33 and 64 mm/yr [22]. Offset streams along strike-slip faults
are often difficult to work with because, by themselves, there is no way to directly
date the offset. At Wallace Creek, it was possible to bracket the slip rate with minimum and maximum age estimates because of the datable stratigraphy associated
with the two stream offsets found there.

Drainage-Pattern Control in Extensional Settings

Active deformation resulting from extension of the crust forms a characteristic
topography of upthrown blocks **(horsts)** and downthrown blocks **(grabens).**
The drainage pattern in such a setting quickly assumes the structural pattern of
the extension. Any throughgoing rivers flow within the grabens. Major tributary streams tend to find their way into the grabens at irregularities in the

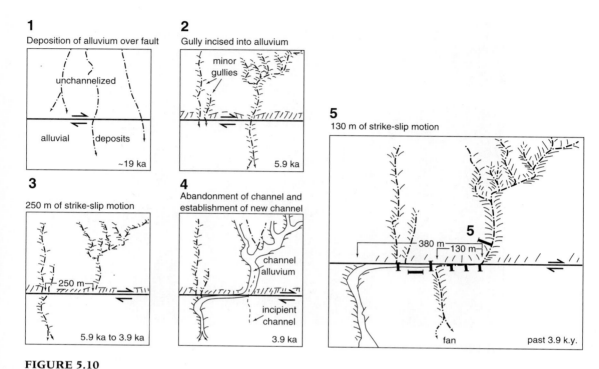

1

Deposition of alluvium over fault

unchannelized

alluvial deposits

~19 ka

2

Gully incised into alluvium

minor gullies

5.9 ka

3

250 m of strike-slip motion

250 m

5.9 ka to 3.9 ka

4

Abandonment of channel and establishment of new channel

channel alluvium

incipient channel

3.9 ka

5

130 m of strike-slip motion

380 m
130 m

5

fan

past 3.9 k.y.

FIGURE 5.10

The history of earthquake surface rupture at Wallace Creek revealed by subsurface trenches.
(From Sieh, K., 1981. In D.W. Simpson and P.G. Richards (eds.), *Earthquake Prediction: An International Review.* American Geophysical Union: Washington, D.C.)

bounding faults, such as fault step-overs [23]. Another common feature of extensional topography is a **half graben**—one of the faults bounding a down-thrown block is dominant, and the other is subsidiary or inactive altogether (Figure 5.11). Through time, offset on the dominant fault will tilt the floor of a half-graben block down in that direction.

Streams in half grabens are affected by active tilting and shift toward the fault [24]. This phenomenon has been recognized in the Hebgen Lake area of Montana, just west of Yellowstone National Park, the site of a magnitude 7.5 earthquake in 1959. The earthquake was accompanied by up to 6.7 m of fault offset and a corresponding amount of down-to-the-northeast tilt on the Hebgen Lake tilt block [25]. Where the Madison River crosses the tilt block, it flows in a broad meander belt, 225–1600 m wide (Figure 5.12). The active channel, however, is located at the very northeast margin of the belt, suggesting that tilt has shifted the channel [26, 27]. However, the asymmetry of the Madison River alone does not reveal the mechanism by which the river shifted. If it suddenly avulsed to one side of the meander belt, such as in a single earthquake, old

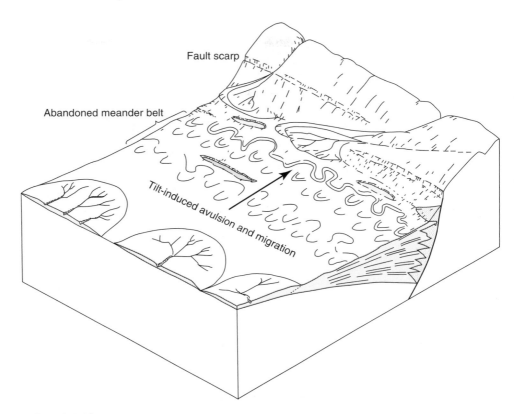

Fault scarp

Abandoned meander belt

Tilt-induced avulsion and migration

FIGURE 5.11
Topography and geomorphology of a half graben. Note that the active channel draining
down the axis of the half graben has shifted toward the active fault scarp by ongoing tilting.
(From Leeder and Gawthorpe, 1987. In M.P. Coward, J.F. Dewey, and P.L. Hancock (eds.), *Continental
Extensional Tectonics*. Geological Society Special Publication, 28)

meander scars would be randomly oriented within the belt (Figure 5.13A). The
meander scars along the Madison River, in contrast, are preferentially oriented,
concave to the northeast (Figure 5.12). This pattern is consistent with gradual
migration of the channel as a result of steady tectonic tilt (Figure 5.13B) that
must predate the 1959 Hebgen Lake earthquake [26, 27].

Lateral migration of streams that run perpendicular to tectonic deformation is
evident elsewhere. Long Valley caldera is a caldera (a volcanic depression) in eastern California formed by an enormous eruption about 750 ka. A portion of the
magma body that caused the eruption remains beneath the caldera, periodically
swelling and uplifting a dome-shaped region [28], most recently causing about 0.5 m
of uplift in 1980. The Owens River (mentioned earlier in this chapter, but about
150 km upstream) runs through Long Valley caldera around the periphery of the

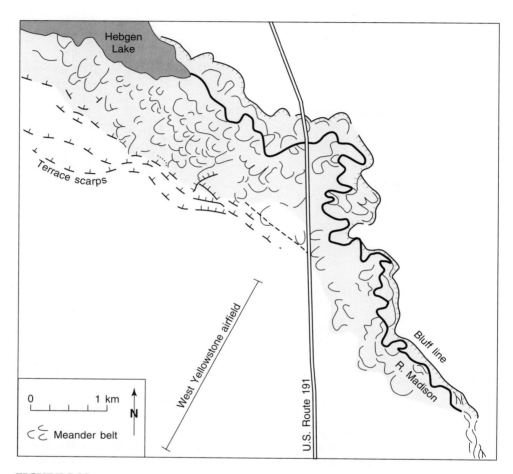

FIGURE 5.12
Geomorphic map of the Madison River in the Hebgen Lake area, just west of Yellowstone
National Park. The active channel lies at the northeast margin of the river's meander belt,
at the base of the bluff line. During the 1959 Hebgen Lake earthquake, this region sub-
sided by as much as 6.7 m down to the northeast.
(From Leeder and Alexander, 1987 [26])

1980 uplift. The river runs in one of two parallel meander belts, one closer to the
zone of uplift and one farther away (Figure 5.14) [29]. Historical records indicate
that the river has been shifting to the downtilt meander belt during the last century
or so [29]. Research in the area concluded that the two belts are maintained by the
periodic nature of uplift in the caldera—the downslope channel is active when the
magma chamber is inflating, and the upslope channel is active when the chamber is
deflating [29].

The same asymmetry in river systems is also seen in ancient sedimentary
deposits and may be a valuable tool to help sedimentologists recognize tectonic

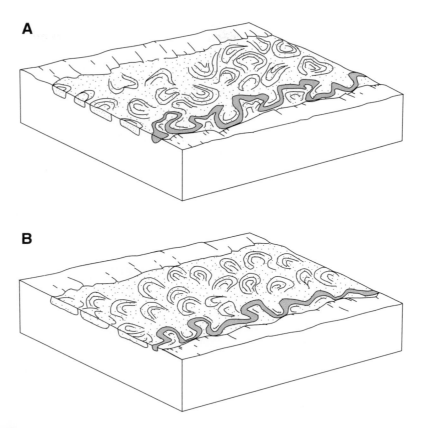

FIGURE 5.13
Models illustrating the effects of tilting perpendicular to a meandering channel. (A) In
the case of a sudden tilting event, the channel shifts with a sudden avulsion and leaves
randomly oriented meander scars. (B) In the case of gradual tilting, the channel shifts
gradually, leaving meander scars preferentially oriented toward the direction of relative
subsidence.
(From Leeder and Alexander, 1987 [26])

deformation active in the geologic past [24]. For example, fluvial deposits found in
northern Wales have the asymmetrical characteristics described above. They reveal
ongoing tilting at the time of accumulation in the Ordovician (about 450 Ma) [30].

INTEGRATED MODELS OF TECTONIC ADJUSTMENT

As the reader is now certainly aware, rivers are highly complex systems.
Researchers have used a variety of approaches to understand the full range of
fluvial responses to active-tectonic deformation, in particular: (1) large sandbox

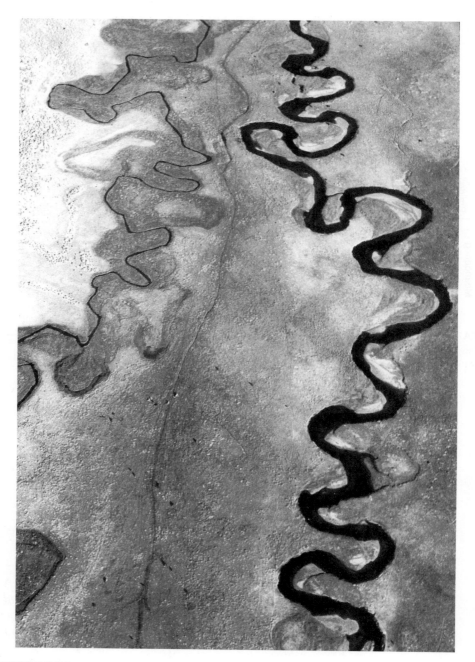

FIGURE 5.14
Parallel meandering channels of the Owens River in Long Valley caldera, looking down toward the northwest. Recent uplift of the resurgent dome of the caldera (left of photograph) has been shifting river flow from the upslope channel to the downslope one. (From Reid, 1992 [29])

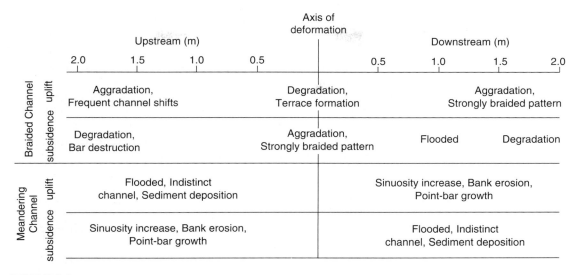

		Upstream (m)				Axis of deformation	Downstream (m)			
		2.0	1.5	1.0	0.5		0.5	1.0	1.5	2.0
Braided Channel	uplift	Aggradation, Frequent channel shifts				Degradation, Terrace formation	Aggradation, Strongly braided pattern			
	subsidence	Degradation, Bar destruction				Aggradation, Strongly braided pattern	Flooded		Degradation	
Meandering Channel	uplift	Flooded, Indistinct channel, Sediment deposition					Sinuosity increase, Bank erosion, Point-bar growth			
	subsidence	Sinuosity increase, Bank erosion, Point-bar growth					Flooded, Indistinct channel, Sediment deposition			

TABLE 5.1
Response of experiment channels to uplift and subsidence across the channel.
(After Ouchi, 1985 [4])

models that re-create fluvial systems in the laboratory, (2) careful observation of rivers in areas of known tectonic activity, and (3) mathematical simulations of river response.

Experimental Models

Ouchi [4] created braided and meandering fluvial patterns in a 9.1-m-long by 2.4-m-wide by 0.6-m-deep flume (a sediment-filled box, inclined and provided with a continuous flow of water from high end to low end). The central 2.4 m of the flume was fitted with a flexible base, allowing incremental raising and lowering of the channel to mimic uplift and subsidence. Four experiments were run using this apparatus:

1. uplift across a braided channel
2. subsidence across a braided channel
3. uplift across a meandering channel
4. subsidence across a meandering channel

The results are summarized in Table 5.1. In experiment 1, the braided channel incised through the axis of uplift, and it aggraded both upstream and downstream of the upwarp. Aggradation downstream was caused by increased sediment load supplied by the incision; aggradation upstream was caused by the reduction in slope. In experiment 2, the effects were opposite those of experiment 1—aggradation across the axis of deformation and degradation upstream and downstream. In experiment 3, the meandering channel adjusted to uplift by changes in sinuosity and flow velocity, with little or no aggradation or degradation. Downstream of the upwarp, sinuosity increased and, in the process, the outer banks of the meanders eroded and point bars inside the meanders grew; upstream of the uplift, flow velocity was reduced and water flooded over the point bars. In experiment 4, the effects were opposite those of experiment 3—sinuosity increased upstream and the current slowed, and flooding occurred downstream [4].

Observational Models

In a very different approach, the effects of deformation were studied by documenting the characteristics of rivers that cross the Monroe and Wiggins uplifts of Louisiana and Mississippi (Figure 5.15). Deformation on these zones is amply documented—Tertiary and Quaternary sedimentary strata are deformed, Pleistocene and Holocene terraces are warped, and geodetic surveys measure ongoing motion [31]. The rivers that cross the zones of uplift show many of the same modifications seen in the flume experiments. Above the axes of uplift, channel sinuosity as

FIGURE 5.15
Locations of the Monroe (M) and Wiggins (W) uplifts.
(After Burnett and Schumm, 1983 [31])

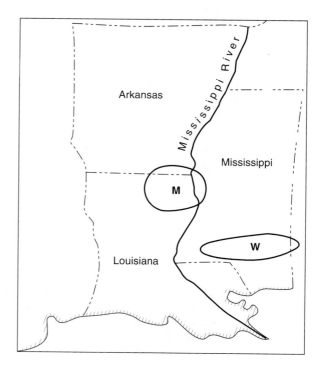

well as channel depth are reduced, and overbank flooding is increased. Below the uplifts, sinuosity and depth are increased, bank erosion is accelerated, and some reaches are locally braided in response to increased sediment loads [31]. In addition, the longitudinal profiles of both the river-valley floors and the channels are convex upward across the Monroe and Wiggins uplifts on all but the largest rivers. The largest rivers apparently have sufficient energy for the channels to cut through the deformation without measurable perturbation [31].

Numerical Models

The dynamics of fluvial systems are sufficiently well known to be modeled on computers. Computer models and laboratory experiments have strengths that real-world studies typically lack; for example, they integrate changes in the system through time rather than just illustrating the characteristics in an instant of geological time (the present). In one such simulation, the effects of several different types of tectonic deformation were modeled for a large river system [32]. Figure 5.16 illustrates how the geometry of fluvial response can change through time. The figure illustrates responses along a 200-km-long river reach across

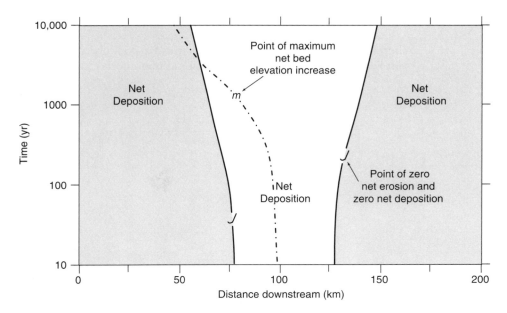

FIGURE 5.16
Zones of aggradation and degradation and critical transition points on a simulated river channel undergoing anticlinal uplift across the channel at the 100-km mark (distance = 0 is upslope). Note that the zones and the transition points gradually shift locations after uplift begins at time = 0.
(After Snow and Slingerland, 1990 [32])

which there is uplift, centered at the 100 km mark, beginning at time = 0. The zone of channel incision widens, and the point of maximum channel-elevation increase (the net effect of uplift and incision) shifts upstream through time. The results of this simulation also emphasize the complexity of such systems. Even though the model was greatly simplified, simple tectonic deformations led to complex river responses, including:

1. reversals of deposition and erosion through time at some points on the channel
2. significant lag times between stimuli and responses
3. critical transition points, such as between zones of aggradation and degradation, that do not coincide with the axes of deformation
4. in general, a mismatch between the location of tectonic deformation and the distribution of fluvial response

Note that the complex responses in this study, sometimes called "nonlinear" responses, are different from the concept of **complex response** introduced in Chapter 2.

RIVER TERRACES

When a river becomes graded, it begins to cut a floodplain (Figure 5.17). The floodplain is a flat strip of land formed by lateral migration of the channel through time.

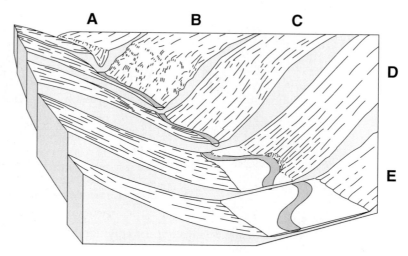

FIGURE 5.17
Transition from disequilibrium river conditions and active incision (A, B, and C) to equilibrium and aggradation of sediment (D and E). The floodplain begins to form after the transition from disequilibrium to equilibrium.

The deposits that comprise a floodplain include channel sediments as well as over-bank deposits from periodic flooding. As discussed previously in this chapter, this sediment substrate is what distinguishes an alluvial river from a bedrock river. If uplift occurs, base level falls, or other factors disrupt the equilibrium of the system, a river may incise through its floodplain in order to reach a new graded profile and begin cutting a new floodplain. In this manner, the old floodplain becomes a river **terrace**—an inactive bench stranded above the new level of the river. Repeated episodes of downcutting may preserve several terraces above a river (Figure 5.18).

River terraces can be classified in terms of the three processes that lead to their formation [33]:

- tectonically-induced downcutting
- climatically-induced aggradation
- complex response

FIGURE 5.18
The Rakaia River in New Zealand. Note the braided channel and floodplain in the fore-ground and the broad, flat terrace across the upper left corner of the photo.
(Photo courtesy of the National Archives of New Zealand, REF. #AAQT 6404 WA 2367)

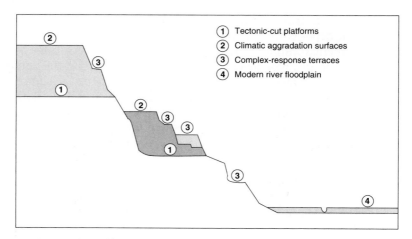

FIGURE 5.19

Classification of tectonic, climatic, and complex-response terraces. Platforms cut onto bedrock reflect periods of tectonic uplift. Aggradation of thick fluvial deposits reflects episodes of climate change. Small unpaired benches reflect critical thresholds and adjustment of internal variables.

(After Bull, 1990 [33])

Terraces in a setting undergoing long-term uplift typically consist of a thin layer of river sediment overlying a bedrock-cut platform (these terraces are called **strath terraces**). Downcutting is characterized by brief periods of equilibrium and floodplain construction separated by periods of incision. Terraces with a thick sedimentary cover (called **fill terraces**) reflect extended periods when resisting forces exceeded driving forces in a system (for example, when there is an abundant supply of sediment). Aggradation of these thick terrace deposits typically is associated with major climatic shifts [33]. Complex response—changes in river systems caused directly by internal variables and thresholds, without tectonic or other external stimuli (see Chapter 2)—forms terraces that are typically small and **unpaired** (isolated, not found on both sides of a river) (Figure 5.19).

Fluvial terraces are probably the landform most widely used to detect and measure tectonic activity from the late Pleistocene to the present. When terrace surfaces become inactive and incised, they do so over short periods of time. As a result, the surfaces become good tools for studying both the distribution and the timing of tectonic deformation. Three types of deformation can be traced across terrace surfaces:

- surface faulting
- warping
- tilting (convergence or divergence of terraces)

Surface Faulting on Terraces

Faults are as likely to cut the surface of a fluvial terrace as any other landform, but several aspects of terraces make them useful for studying faulting. Terraces are quite flat, so fault scarps are preserved longer than they would be on a steep slope. Furthermore, terrace surfaces represent approximate geomorphic time-lines. If a fault cuts a given terrace, then faulting is known to *postdate* that surface; if a fault crosses but does not cut a given terrace, then faulting must *predate* that surface. Finally, the age of fault scarps cut entirely in loose sediment some-times can be estimated directly by **diffusion-equation modeling,** a process that is outlined in Chapter 8 (Fault-Scarp Degradation).

Many studies have recognized faulting on river terraces [e.g., 34, 35, 36]. In one case, initial field mapping recognized a large number of different surfaces, but careful remapping and correlation of surfaces using soil ages revealed that there were only four terrace levels [34]. The four terraces were cut and offset by a number of fault strands. By knowing the ages of the different terraces and mea-suring the surface displacements, it was possible to calculate the rate of slip on the fault system [34].

Warping of Terraces

Terrace surfaces are formed as river floodplains. Because they are in equilibrium when formed, terraces have graded longitudinal profiles unless they have been sub-sequently deformed. Deformation will alter the profiles of terraces compared with the longitudinal profile of the modern channel. The shapes of deformed terraces reveal the character of deformation (Figure 5.20C). An assumption implicit in this comparison is that the original terrace profile was the same as the profile of the modern river. Changes in discharge, sediment load, or bedrock substrate strength may invalidate this assumption. An important test of tectonic deformation is whether higher and older terraces are more deformed than younger terraces. Cli-matically or lithologically controlled variations in profile are unlikely to progress systematically through time, although they may do so in certain situations.

Longitudinal profiles of terraces have been used to recognize recent defor-mation in many studies [e.g., 36, 37, 38]. In the heart of the Los Angeles metro-politan area, Pleistocene and Holocene terraces and broad geomorphic surfaces are warped over two active anticlines, revealing as much as 150 m of uplift and up to 30° of tilting [36]. These anticlines probably represent the surface manifes-tations of buried thrust faults similar to the one that caused the 1994 Northridge earthquake (see Chapter 7).

Tilting of Terraces

Simple downcutting on a river system without any contemporaneous deformation will lead to a sequence of parallel terraces. Where downcutting is accompanied by

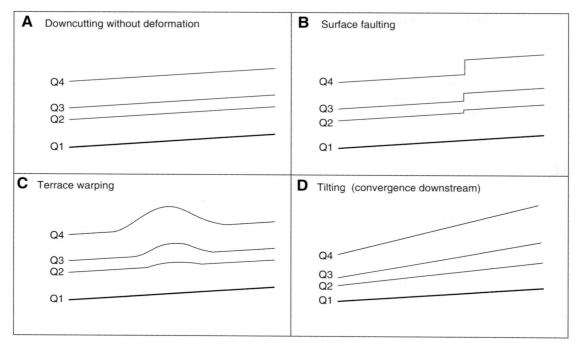

FIGURE 5.20
Four types of tectonic deformation of fluvial terraces: (A) uplift and incision without differential deformation, (B) surface faulting, (C) terrace warping, and (D) tilting. Note that cases B, C, and D illustrate progressive deformation through time—the oldest terraces are most deformed.

regional-scale tilting, however, longitudinal terrace profiles will diverge or converge (Figure 5.20D). As with terrace warping, other processes can cause divergence or convergence; progressive displacement of a series of terraces is an important test before tectonic deformation can be inferred.

SUMMARY

Rivers are profoundly influenced by active tectonics. Fluvial systems are sensitive to both faulting and regional surface deformation. As a result, it is possible to use perturbations in these systems to locate, characterize, and measure recent tectonic activity. This type of research is a relatively recent development in geomorphology, and the study of rivers and active tectonics continues to evolve. Several avenues of research are now being developed that may lead to great advances in the future. Digital elevation models and a variety of satellite images make recognition of fluvial patterns increasingly straightforward [5]; many parameters can even be measured by automatic computer analysis [39]. In addition, advances in dating techniques [40, 41] should increase opportunities for dating terraces and other landforms, widening the range of settings from which tectonic rates can be calculated. Finally, we look forward to the full integration of fluvial geomorphology and tectonic geomorphology, with the recognition of tectonic effects in many fluvial systems and the recognition that rivers present valuable tools for use in studies of active tectonics.

REFERENCES CITED

1. Bull, W.B., 1991. *Geomorphic Response to Climate Change.* Oxford University Press: New York.
2. McLennan, S.M., 1993. Weathering and global denudation. *Journal of Geology,* 101: 295–303.
3. Adams, J., 1980. Active tilting of the United States midcontinent: geodetic and geomorphic evidence. *Geology,* 8: 442–446.
4. Ouchi, S., 1985. Response of alluvial rivers to slow active tectonic movement. *Geological Society of America Bulletin,* 96: 504–515.
5. Deffontaines, B., and J. Chorowicz, 1991. Principles of drainage basin analysis from multisource data: application to the structural analysis of the Zaire Basin. *Tectonophysics,* 194: 237–263.
6. Whitney, J.D., 1872. The Owens Valley earthquake. *The Overland Monthly,* 9: 130–140, 266–278.
7. Lyell, C., 1857. *Principles of Geology.* Appleton and Company: New York. (cited in S.A. Schumm, 1986. Alluvial river response to active tectonics. In *Active Tectonics: Studies in Geophysics.* National Academy Press: Washington, D.C.)
8. Saucier, R.T., 1987. Geomorphological interpretation of late Quaternary terraces in western Tennessee and their regional tectonic implications. In D.P. Russ and A.J. Crone (eds.), The New Madrid, Missouri, Earthquake Region—Geological, Seismological, and Geotechnical Studies. *U.S. Geological Survey Professional Paper* 1336-A.
9. Schumm, S.A., 1986. Alluvial river response to active tectonics. In *Active Tectonics: Studies in Geophysics.* National Academy Press: Washington, D.C.
10. Mackin, J.H., 1948. Concept of the graded river. *Geological Society of America Bulletin,* 59: 463–512.
11. Leopold, L.B., and T. Maddock, 1953. Hydraulic geometry of stream channels and some physiographic implications. *U.S. Geological Survey Professional Paper* 252.
12. Rhea, S., 1989. Evidence for uplift near Charleston, South Carolina. *Geology,* 17: 311–315.
13. Marple, R.T., and P. Talwani, 1993. Evidence of possible tectonic upwarping along the South Carolina coastal plain from an examination of river morphology and elevation data. *Geology,* 21: 651–654.
14. Berberian, M., M. Qorashi, J.A. Jackson, K. Priestley, and T. Wallace, 1992. The Rudbar-Tarom earthquake of 20 June in NW Persia: preliminary field and seismological observations, and its tectonic significance. *Bulletin of the Seismological Society of America,* 82: 1726–1755.
15. Gomez, B., and D.C. Marron, 1991. Neotectonic effects on sinuosity and channel migration, Belle Fourche river, western South Dakota. *Earth Surface Processes and Landforms,* 16: 227–235.
16. Merritts, D., and K.R. Vincent, 1989. Geomorphic response of coastal streams to low, intermediate and high rates of uplift, Mendocino triple junction region, northern California. *Geological Society of America Bulletin,* 101: 1373–1388.

17. Merritts, D., and W.B. Bull, 1989. Interpreting Quaternary uplift rates at the Mendocino triple junction, northern California, from uplifted marine terraces. *Geology,* 17: 1020–1024.

18. Abrahams, A.D., 1984. Channel networks: a geomorphological perspective. *Water Resources Research,* 20: 161–188.

19. Engebretson, D.C., A. Cox, and R.A. Gordon, 1985. Relative motion between oceanic and continental plates in the Pacific basin. *Geological Society of America Special Paper* 205.

20. Wood, L.J., F.G. Etheridge, and S.A. Schumm, 1990. Effects of base-level change on coastal plain–shelf slope systems: an experimental approach. *American Association of Petroleum Geologists Bulletin,* 74: 1349.

21. Schumm, S.A., 1993. River response to baselevel change: implications for sequence stratigraphy. *Journal of Geology,* 101: 279–294.

22. Sieh, K., and R.H. Jahns, 1984. Holocene activity of the San Andreas fault at Wallace Creek, California. *Geological Society of America Bulletin,* 95: 883–896.

23. Paton, S., 1992. Active normal faulting, drainage patterns, and sedimentation in southwestern Turkey. *Journal of the Geological Society, London,* 149: 1031–1044.

24. Leeder, M.R., and R.L. Gawthorpe, 1987. Sedimentary models for extensional tilt-block/half-graben basins. In M.P. Coward, J.F. Dewey, and P.L. Hancock (eds.), Continental Extensional Tectonics. *Geological Society Special Publication,* 28: 139–152.

25. Myers, W.B., and W. Hamilton, 1964. Deformation associated with the Hebgen Lake earthquake of August 17, 1959. *U.S. Geological Survey Professional Paper* 435: 55–98.

26. Leeder, M.R., and J. Alexander, 1987. The origin and tectonic significance of asymmetrical meander-belts. *Sedimentology,* 34: 217–226.

27. Alexander, J., and M.R. Leeder, 1990. Geomorphology and surface tilting in an active extensional basin, SW Montana, U.S.A. *Journal of the Geological Society, London,* 147: 461–467.

28. Bailey, R.A., G.B. Dalrymple, and M.A. Lanphere, 1976. Volcanism, structure, and geochronology of Long Valley Caldera, Mono County, California. *Journal of Geophysical Research,* 81: 725–744.

29. Reid Jr., J.B., 1992. The Owens River as a tiltmeter for Long Valley caldera, California. *Journal of Geology,* 100: 353–363.

30. Orton, G.J., 1991. Emergence of subaqueous depositional environments in advance of a major ignimbrite eruption, Cape Curig Volcanic Formation, Ordovician, North Wales—an example of regional volcanotectonic uplift? *Sedimentary Geology,* 74: 251–286.

31. Burnett, A.W., and S.A. Schumm, 1983. Alluvial-river response to neotectonic deformation in Louisiana and Mississippi. *Science,* 222: 49–50.

32. Snow, R.S., and R.L. Slingerland, 1990. Stream profile adjustment to crustal warping: nonlinear results from a simple model. *Journal of Geology,* 98: 699–708.

33. Bull, W.B., 1990. Stream-terrace genesis: implications for soil development. *Geomorphology,* 3: 351–367.

34. Rockwell, T.K., E.A. Keller, M.N. Clark, and D.L. Johnson, 1984. Chronology and rates of faulting of Ventura River terraces, California. *Geological Society of America Bulletin,* 95: 1466–1474.

35. Bishop, P., and J.C. Bousquet, 1989. The Quaternary terraces of the Lergue River and activity of the Cévennes fault in the lower Hérault Valley (Languedoc), southern France. *Zeitscrift fur Geomorphologie,* 33: 405–415.

36. Bullard, T.F., and W.R. Lettis, 1993. Quaternary fold deformation associated with blind thrust faulting, Los Angeles Basin, California. *Journal of Geophysical Research,* 98: 8349–8369.

37. Reheis, M.C., 1985. Evidence for Quaternary tectonism in the northern Bighorn Basin, Wyoming and Montana. *Geology,* 13: 364–367.

38. Pinter, N., and E.A. Keller, 1992. Quaternary tectonic and topographic evolution of the northern Owens Valley. In C.A. Hall Jr.,

V. Doyle-Jones, and B. Widawski (eds.), *The History of Water: Eastern Sierra Nevada, Owens Valley, White-Inyo Mountains*. White Mountain Research Station: Los Angeles.

39. Gardner, T.W., K.C. Sasowsky, and R.L. Day, 1990. Automated extraction of geomorphometric properties from digital elevation data. *Zeitscrift für Geomorphologie Supplement,* 80: 57–68.

40. Cerling, T.E., 1990. Dating geomorphic surfaces using cosmogenic ^{3}He. *Quaternary Research,* 33: 148–156.

41. Zreda, M.G., F.M. Phillips, P.W. Kubik, P. Sharma, and D. Elmore, 1993. Cosmogenic ^{36}Cl dating of a young basaltic eruption complex, Lathrop Wells, Nevada. *Geology,* 21: 57–60.

6

Active Tectonics and Coastlines

INTRODUCTION

On April 25, 1992, a $M = 7.1$ earthquake struck the northern California coast near Cape Mendocino (Figure 6.1), causing moderate damage to small towns in the area. The earthquake occurred along the Cascadia Subduction Zone, where the Juan de Fuca Plate subducts beneath the North American Plate. In the days and weeks after the earthquake, subtle changes to the coastline of northern California became apparent. The first accounts came from fishermen and residents of the coast near the epicenter who reported changes in familiar offshore rocks and reefs. Within one to two weeks, attached marine organisms began to

FIGURE 6.1
Location map of the Cascadia subduction zone. The star indicates the site of the
April 25, 1992, earthquake near Cape Mendocino. The Juan de Fuca and the Explorer
plates are being carried beneath the North American Plate. In contrast, where the Pacific
Plate is in direct contact with the North American Plate, south of Cape Mendocino and
north of the Explorer Plate, the plate boundaries are transform faults.

die at the upper margin of the intertidal zone, forming a stripe of bleached-out
death along 23 km of the shoreline (Figure 6.2). Scientists who investigated the
natural destruction determined that the earthquake had uplifted the coast by
up to 1.4 m, carrying a whole intertidal ecosystem up and out of the ocean [1].

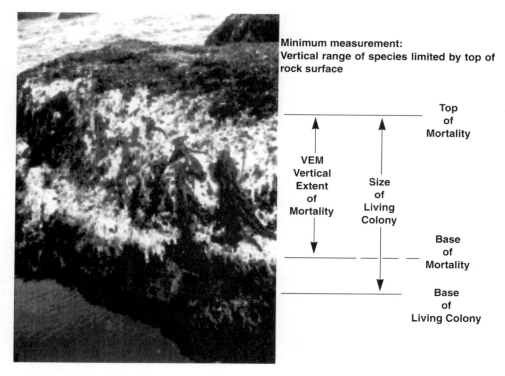

Minimum measurement:
Vertical range of species limited by top of
rock surface

Top
of
Mortality

VEM
Vertical
Extent
of
Mortality

Size
of
Living
Colony

Base
of
Mortality

Base
of
Living Colony

FIGURE 6.2
Photograph of the mortality of intertidal organisms caused by the 1992 Cape Mendocino
earthquake.
(From Carver et al., 1994 [1])

While such events are rare in the timescale of human events, they are common
in geologic time.

Tectonic activity occurs in all settings of the planet, from the deepest ocean
trench to the highest mountain peak. But activity that coincides with ocean or
lake coastlines has unique effects, and the record of tectonic activity at coasts
contains data of great value in determining past earthquake history. The ocean
surface is Earth's best approximation of a perfect horizontal datum (see discussion of the **geiod** in Chapter 3). Apart from tides and variations in atmospheric
pressure, which average out over time, sea level precisely follows a surface of
equal-gravity potential everywhere around the Earth. Any change of position
relative to sea level disrupts the delicate balance of biological, chemical, and
physical forces concentrated there.

Coastal landforms can be used to decode several aspects of active tectonics.
After the April 1992, Cape Mendocino earthquake, scientists used the height of
the "kill zone" of intertidal organisms to determine the amount of coseismic uplift
(up to 1.4 m) and the pattern of the uplift (an east-west–trending arch) [1].

Although these "kill zones" are useful for measuring coseismic uplift, they are not as useful for measuring gradual, long-term uplift or deformation of coastlines, simply because the biological evidence is not preserved longer than a few months to years. Long-term uplift is recorded by more durable features, such as marine terraces, beach ridges, and coral reefs. Coastlines are further useful for studies of active tectonics because many coastal landforms act as effective isochrons (time-lines) and allow us to estimate the age of faulting or other deformation events.

The character of a coastline is a function of many variables, but four stand out:

1. the local tectonic setting
2. the supply of sediment
3. the amount of erosional energy
4. the nature and amount of biological activity

The rugged, cliff-lined coasts of California, Oregon, and Washington are distinct from the sandy, barrier-island coastlines of the East Coast of the United States because, in much of the West, the land is steadily rising relative to sea level. In addition, the Pacific Coast is starved of sand compared to the Atlantic, and the energy of breaking waves is focused onto the cliffs. Figure 6.3 illustrates the classification of coastlines along two axes, from **emergent** (rising relative to sea level) to **submergent** (falling relative to sea level), and from **retrograded** (retreating landward through time) to **prograded** (advancing oceanward through time).

Coastal landforms are profoundly influenced by changes in sea level relative to the land. Since the maximum of the last Ice Age, about 18 ka, melting glaciers and warming oceans have raised global sea level by about 125 m, drowning continental shelves worldwide and creating coastal landforms intimately associated with rising sea levels. Among the stable to emergent coastlines, there are many different examples of retrograded and prograded shorelines. The Pacific Coast of the United States is emergent, the Gulf of Mexico near the Mississippi River is submergent, and the Atlantic Coast is close to neutral. In this chapter, three classes of coasts are discussed:

- **erosional** ("Pacific-type": high erosional energy relative to sediment supply)
- **clastic** ("Atlantic-type": high sediment supply relative to erosional energy)
- **carbonate** (coasts characterized by the formation of coral reefs)

These three classes are associated with different assemblages of landforms, each recording the occurrence, character, and rate of active tectonic deformation in a given region.

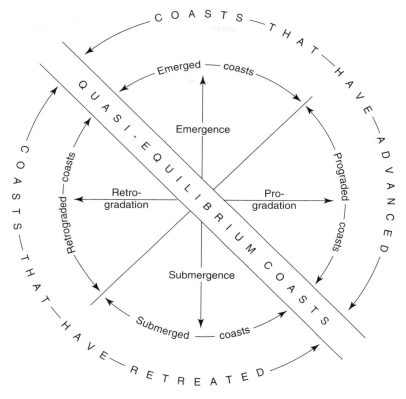

FIGURE 6.3

Genetic classification of coastlines. Coasts are categorized along two axes: from "emerged" to "submerged," and from "prograded" to "retrograded."

(After Valentin, 1970. Paper read at the Symposium of the IGU Commission on Coastal Geomorphology: Moscow)

COASTAL LANDFORMS

Erosional Coasts

Bold erosional coasts have a characteristic morphology (Figure 6.4). At low tide, one can walk along the **wave-cut platform,** a wide and flat surface cut by the abrasive action of wave-churned sand, chemical weathering of the rock by seawater, and the action of intertidal organisms. Where organisms are the principal agent of erosion, such as in tropical shorelines protected from wave action, the cut platform is called a **bioerosion platform** [2]. At the landward margin of the platform is the **seacliff.** The seacliff is a geomorphic anomaly—the steep face is maintained only by the perpetual battering of high-tide and storm waves at its

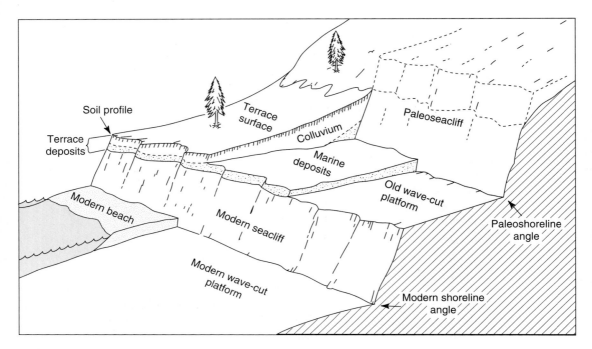

FIGURE 6.4
General morphology of an erosional coastline. Active features include the modern wave-cut platform, beach, and seacliff. An older, uplifted, and now inactive shoreline is recorded by the old wave-cut platform, its cover of marine sediment, and the eroded remnant of the old seacliff. Note that the shoreline angles, both modern and uplifted, are the only uniform reference to sea level.
(After Hanson et al., 1990. *Guidebook, Friends of the Pleistocene Fall Field Trip*)

base. Wave erosion of a durable rock may form a **wave-cut notch** at the base of the cliff, an overhung depression that is a good indicator of sea level. Long-term seacliff retreat makes clifftop homes scenic but ephemeral real estate [3].

When a wave-cut platform is uplifted above sea level, the landform becomes known as an **uplifted marine terrace** (see Figure 6.4). The surface of the old platform may, with luck, still contain fossils of its previous inhabitants, and generally will be covered with a veneer of loose marine sediments. The uplifted seacliff is no longer maintained by wave erosion, so it will degrade. Through time, the marine terrace becomes covered by colluvium from the old seacliff and by wash from farther upslope.

It is important to note that the top surfaces of active wave-cut platforms and uplifted marine terraces typically have a natural upward slope away from the ocean. For that reason, neither the top edge of a seacliff nor any random point on the surface is a good estimate of the sea level that formed the terrace. The only uniform horizontal reference to sea level on an uplifted marine terrace

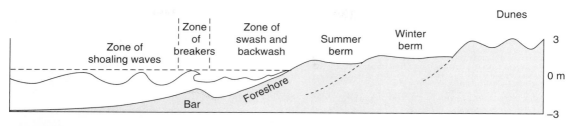

FIGURE 6.5
General morphology of a clastic shoreline.
(After Easterbrook, 1993 [32])

is the **shoreline angle** (see Figure 6.4), which is the angle between the old platform and the associated seacliff.

Clastic Coasts

Sandy barrier islands from Fire Island, New York, to Chincoteague, Maryland, to the Outer Banks of North Carolina, to Padre Island, Texas, are all examples of clastic coastlines. All are associated with ample sediment supply and shore-parallel transport (**longshore transport**) of sand along the beach. The particular nature of barrier islands, separated from the mainland by broad lagoons, is attributed to rising postglacial sea level [4]. Clastic coasts are associated with a variety of landforms, but the shore has a generalized profile (Figure 6.5) that reflects the fundamental processes of wave action and sediment transport.

Of the features of the shoreface profile, the **winter berm** (or **storm berm**) is the "high-water mark" of a given shoreline. More specifically, the berm is a topographically-high, laterally-continuous feature that may be preserved if the active shoreline shifts oceanward. Such a shift can be caused by (1) an influx of sediment and progradation of the coast (Figure 6.6), (2) a relative fall in sea level (see discussion of sea level later in this chapter), or (3) uplift—either coseismic or gradual—of the land (Figure 6.7).

Carbonate Coasts

Coral reefs are constructed by biological activity in warm (>18°C), clear, oxygenated water (Figure 6.8) [4]. Modern theories of coral reef development can be traced back to Charles Darwin, who made many observations during the voyage of the *Beagle* from 1832 to 1835. He characterized different reefs around the Pacific, and suggested a hypothesis to explain their origin by subsidence of extinct island volcanoes and upward growth of the coral.

The growth of coral is closely tied to sea level (Figure 6.9). Active colonies can grow only as high as the lowest low tide. Reef corals live no deeper than the

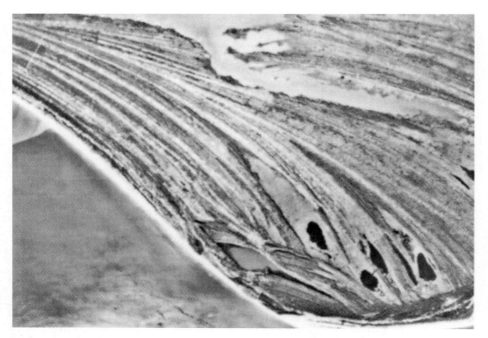

FIGURE 6.6
Multiple prograded beach ridges developed at St. Vincent Island, Florida. Prograded beach
ridges are formed by a massive influx of sediment, and not necessarily by either uplift or
sea-level change.
(Photo courtesy of U.S. Geological Survey)

depth that light penetrates through the water. Most species live at depths of less
than 25 m, with maximum growth shallower than 10 m [5]. Coral reefs are a
complex ecosystem, and particular species—for example, *Acropora palmata*—pre-
dominate at shallow depths. In fossil reefs, these species are the best indicators
of the position of sea level when the reef formed.

Some of the best-documented examples of tectonic coastlines contain uplifted
coral reefs [6, 7]. Reefs have some characteristics that make them ideal for studying
active tectonics. First, uplifted coral reef surfaces are good time lines, because the
first vertical movement carrying the reef crest out of water kills the organisms and
preserves the surface as a record of sea level at that moment. Second, the carbonate
material of which coral reefs are constructed usually can be dated. Radiocarbon and
uranium-series analyses are used to date uplifted reefs as old as 350 ka (for more
details, see Dating Coastal Landforms at the end of this chapter).

COSEISMIC DEFORMATION

Many of the large and great earthquakes of historical time have been accompanied
by measurable deformation of the surface. In 1964, approximately 250,000 km² of

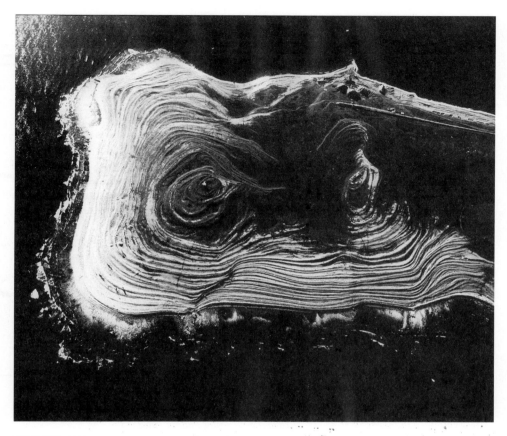

FIGURE 6.7
Emerged Holocene beach ridges on Östergarnsholm, Sweden. Ongoing uplift is driven by isostatic compensation for the melting of the ice cap that covered Scandinavia during the Pleistocene. Although the ice has been gone for several thousand years, isostatic compensation continues and causes very rapid uplift, at a rate of about 2 m/k.y. at this location. (Photograph courtesy of A. Philip, Visby, Sweden)

the Alaskan coast experienced vertical deformation, with uplift of up to 10 m and subsidence of up to 2.4 m (see Figure 1.20) [8]. Such episodic pulses of uplift, summed over the eons of geological time, are one mechanism by which the large-scale topography of the Earth is created (see also Chapter 9). When earthquakes strike near a coastline, it is often possible to measure the magnitude and pattern of the coseismic deformation by virtue of the nature of coastal landforms and their link to the global sea-level datum. Different types of coastlines provide a range of tools for studying coseismic vertical deformation:

- tide-gauge data
- mortality of intertidal organisms ("kill zones")

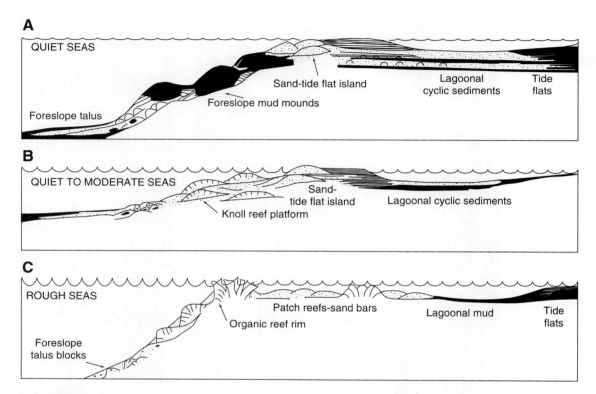

FIGURE 6.8
Three types of carbonate coastlines.
(After Wilson, 1974. *American Association of Petroleum Geologists Bulletin,* 58: 811)

- beach strandlines
- submerged stratigraphy

Tide Gauges

It may be a bit counterintuitive, but the surface of an ocean is a very precise horizontal level. Indeed, at any one time, the water is tossed by waves, drawn in and out with the tides, and pushed about by currents and weather. However, when all of these variations are averaged *over a long period of time,* the result is a fixed elevation accurate to within millimeters. Where monitoring stations are set up to systematically measure the average level of the ocean surface at a point on the coast, at ports or at scientific stations, it then becomes possible to precisely determine the elevation of that point through time relative to sea level (Figure 6.10). These monitoring stations are known as **tide gauges,** and 517 stations world-

FIGURE 6.9
A coral reef at low tide. Note that the upper limit of growth is approximately equal to the low-tide water level.
(Photo by Douglas Faulkner/Photo Researchers, Inc.)

wide have records from a long enough duration of time and have proven sufficiently stable to provide geodetic-quality measurements of mean sea level [9].

Where an earthquake causes coseismic deformation near one of these tide gauges, the amount of vertical movement is recorded as an abrupt change in the elevation of the land relative to mean sea level. In fact, coseismic movement is one of the most straightforward applications of tide-gauge data. In addition, by 1993, approximately 125 of the stations had records spanning 52 years or more, a duration of time long enough, given the data's accuracy, to measure long-term changes in global sea level as well as long-term uplift of the land due to subduction, glacial unloading, and volcanism [9]. Tide gauges are especially valuable in documenting deformation because they are a continuous record of local sea level through time, capable of distinguishing between preseismic, coseismic, and postseismic movements (see Chapter 3).

There are, however, drawbacks to tide-gauge data. Coseismic deformation is local and variable in magnitude. Because tide gauges measure movement only at a few scattered points, such measurements (1) are only minimum estimates of the maximum uplift magnitude, and (2) provide little information regarding the *pattern* or distribution of uplift.

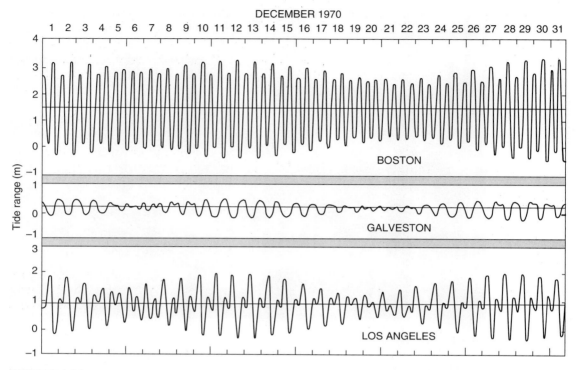

FIGURE 6.10
Tidal curves for three stations along the coasts of the United States. Note that although the
water level is quite variable, the average sea level can be determined with great accuracy.
(After Emery and Aubrey, 1991 [9])

Mortality of Intertidal Organisms

The uplift and death of organisms, such as after the 1992 event at Cape Mendocino described previously, have also been observed after other subduction-zone
earthquakes. Such kill zones have been used to measure the coseismic uplift
associated with the 1964 Alaska earthquake [10], the 1985 Mexico City earthquake [11], the 1960 Chile earthquake [12], the 1992 $M = 7.2$ earthquake in
New Guinea [13], and others. The principal advantage of studying the mortality
of intertidal organisms is that data are available over a wide area—wherever
uplift coincides with the coastline.

Looking at the kill zone along the Mexican coast after the 1985 earthquake, Bodin and Klinger [11] defined the **vertical extent of mortality**
(VEM) as the average height of mortality of attached intertidal organisms, measured by surveying equipment. Just as tide gauges average the short-term fluctuations of the ocean surface, organisms living in the intertidal zone are adapted to

a particular position relative to sea level—exposed to air no more nor less than a certain portion of the time, and just out of reach of their fiercest predators. For example, mussels form a dense band along the intertidal zone of some rocky shores, a band with abrupt upper and lower boundaries. Any mussel that takes a position too high is stranded out of water too long and dies; any mussel that is too low quickly falls victim to hungry starfish inhabiting the subtidal zone below. Organisms that prove particularly useful in measuring VEM include species of coralline algae, barnacles, mussels, colonial anemones, surf grass, and sea lettuce [1]. After the Mendocino earthquake, mortality was evident within one to two weeks. Measurements taken after one month and again two months after the earthquake showed a slight increase in the VEM, but little more than the range-of-measurement uncertainty [1].

Beach Strandlines

Coseismic uplift of a clastic coast will tend to shift the beach profile (see Figure 6.5) upward and out of equilibrium with the forces that formed it. On the lower portion of the profile, this shift brings the zone of wave action closer and erodes material to re-equilibrate the shape of the shore. However, the upper portion of the profile, particularly the high winter-storm berm, is removed from wave erosion and may be preserved as a **beach ridge,** landward of and higher than the active berm that reforms. Inactive berms also can be formed by oceanward progradation of the shore (see Figure 6.6), but these beach ridges will be at approximately the same elevation as the active berm.

Late Holocene beach ridges that step upward and landward from the active shoreline are characteristic of many tectonically active, clastic coasts. An important question about a given beach-ridge sequence is: Why does it have a particular number and spacing of ridges? Three mechanisms are possible: (1) small-scale fluctuations in Holocene sea level, (2) gradual uplift of the land, punctuated by severe storms, and (3) episodic coseismic uplifts. Given the general lack of global synchroneity in beach ridges and the lack of independent evidence of Holocene sea-level fluctuations, many scientists reject the first explanation and invoke either the second or the third [14].

Beach ridges often are difficult to date, usually providing little information on the precise dates of past earthquakes. However, if the ridges formed coseismically, they may provide good information regarding the number of events and earthquake recurrence over time periods of several thousand years. On the Boso Peninsula of Japan, an earthquake in 1703 lifted the active shoreline up and out of contact with the ocean. Three other paleoshorelines step upward from the pre-1703 shore, the highest dated at about 6.5 ka [15]. Those landforms reveal that at least four coseismic uplift events of the magnitude of the 1703 earthquake have occurred on the Boso Peninsula in the past 6500 years.

Coseismic Subsidence

In most cases, downward movement of the land relative to the ocean is much more difficult to discern than upward movement. Subsidence covers the coastal landscape with water and sediment, hiding it from the scrutiny of air-breathing geologists.

Along the coast of Washington and Oregon, at the convergent boundary between the small Juan de Fuca Plate and the North American Plate, salt marshes and coastal forests display an unusual stratigraphy (Figure 6.11). Long periods of gradual sedimentation and slow emergence from the ocean led to the formation of stable coastal landforms capped by soil and vegetation. These stable surfaces were resubmerged beneath sea level, apparently catastrophically. During the resubmergence, marsh grasses were buried either by intertidal mud or, in some locations, by a layer of coarse sand interpreted as a tsunami deposit [16, 17]. Coastal forests were inundated and the stumps, still rooted in the soil from which they grew, were buried by marine sediments. Examination of the outermost growth rings of these trees shows that growth was normal during the last few years before the death of the trees, suggesting that submergence was sudden [16]. In each case, submergence was followed by renewed emergence from the ocean, soil formation, and vegetation growth.

The stratigraphy of coastal Oregon and Washington is interpreted to reflect sudden, coseismic subsidence during great earthquakes on the Cascadia Subduction Zone [18]. Although this area has experienced no great earthquake ($M \geq 8$) during the two centuries of European occupation, some geologists infer that the coseismic subsidence events record thrust-fault earthquakes on the subduction zone at least as large as $M = 7.6$–7.8 [19], perhaps even rivaling the $M = 9.5$ 1960 Chile earthquake (largest recorded in history) if the entire 1200 km length of the Cascadia subduction zone were to rupture all at once [19].

As many as six different coseismic subsidence events along the coast have been suggested [20]. High-precision radiocarbon dating suggests that events at widely separated sites on the coast may have been simultaneous [18]. In particular, evidence is widespread that one earthquake occurred about 300 yr ago, probably between 1695 and 1710 [18, 19, 20, 21].

Some scientists doubt that the stratigraphic and other data have been correctly interpreted as evidence of great earthquakes along the Cascadia subduction zone. It has been pointed out that for fourteen great earthquakes measured on nine different subduction zones around the world, areas of coasts in the same setting as most of coastal Oregon and Washington experienced *uplift* rather than subsidence [22]. Submerged stratigraphy could reflect the effects of several much smaller earthquakes, rather than one great one. Alternatively, it is possible that the evidence reflects only very large storm events superimposed upon the much slower processes of sediment accumulation, Holocene sea-level rise, and episodes of gradual uplift or subsidence. Nonetheless, the consensus is that the Oregon-Washington coast does record

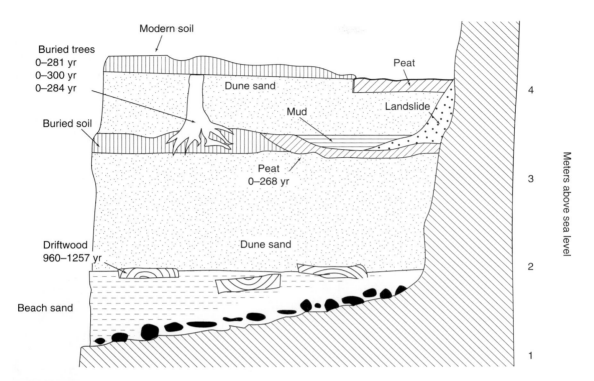

FIGURE 6.11
Schematic cross section through Holocene salt-marsh deposits along the Cascadia subduction zone. The dune sand is interpreted to represent two coseismic subsidence events in the past 960–1257 yr. Ages in years before present.
(After Clarke and Carver, 1992 [19])

long periods of stability punctuated by large earthquakes and regional coseismic subsidence.

COASTAL GEOMORPHOLOGY AND SEA LEVEL

No discussion of coastal geomorphology or movement of the land relative to the ocean can be separated from the dramatic fluctuations in sea level that have occurred during the last 2 m.y. or so. Changes in sea level (**eustatic** change) during this interval primarily reflect the growth and decay of the great ice sheets that repeatedly covered much of North America and Europe. These ice sheets reached thicknesses of up to 3000 m [4], storing up to 100 billion m^3 of water [23] on the continents and depleting the world's oceans proportionally. At the time of the last glacial maximum, about 18 ka, global sea level was approximately 125 m lower than it is today [6].

Coastal landforms can only record changes in the position of the land *relative to* the position of sea level. The number of coastal features, their spacing, and their character are a function of both tectonic movement and sea-level change. In order to say anything about the history of tectonics, we must first know something about the history of sea level.

The best records of sea level come not from stable coastlines, but from coasts experiencing steady long-term uplift. Sea-level change on stable coastlines repeated itself during the glacial advances and interglacial retreats of the Pleistocene ice sheets, destroying evidence of older highs and lows. On uplifting coasts, though, shorelines were progressively lifted out of the range of eustatic change. Sea-level data have been collected around the world, but the two most complete records are in Barbados [7] and New Guinea [6] (Figure 6.12). At both of these sites, coral reefs formed at approximately the times of sea-level

FIGURE 6.12
Uplifted coral reef terraces at the Huon Peninsula, New Guinea. Relatively rapid fluctuations of global sea level during the Pleistocene have been superimposed on long-term tectonic uplift of the New Guinea coast. Each terrace indicates a former position of the shore. Because coral can be dated, each terrace also gives the date at which the shore was at that position.
(Photograph courtesy of A.L. Bloom)

highstands and were uplifted by long-term tectonic activity. Dating the carbonate in each uplifted reef reveals a history of sea-level highs at 60 ka, 82 ka, 105 ka, 125 ka, 170 ka, and 230 ka in Barbados [7] and at 28 ka, 40 ka, 60 ka, 82 ka, 103 ka, and 124 ka in New Guinea [6]. The agreement between these two sites, located at completely different areas of the globe, has led to widespread acceptance of the nature and timing of global sea-level change.

In addition to the timing of past sea-level events, it is important to know the elevations of the different highstands and lowstands. In Barbados and New Guinea, this information was obtained by calculating the long-term uplift rate. For a reef or other feature formed at sea level, its present elevation is equal to

$$\text{elevation} = (\text{uplift rate}) \bullet \text{time} - (\text{sea level}) \tag{6.1}$$

where *time* refers to the age of the feature and *sea level* to the level of the ocean surface at that time, measured in distance below modern sea level. By assuming that the sea level of the last interglacial reef (125 ka) was about 6 m higher than modern sea level (an assumption supported by data from stable, nontectonic coasts), it became possible to calculate the uplift rates for the sites at Barbados and New Guinea and then to determine the position of sea level at the time of the formation of each reef. A summary of the New Guinea data is shown in Figure 6.13,

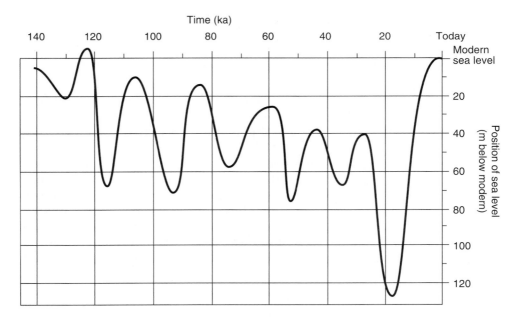

FIGURE 6.13
A graphical representation of global sea level during the past 140 ka. This sea-level model utilizes the data from New Guinea coral reefs (Bloom et al., 1974 [6]) and an analysis technique (Modified after Pinter and Gardner, 1989 [25]). The formula for calculating this model numerically is outlined in Table 6.1.

TABLE 6.1
Sea level (SL) is calculated as follows: $SL = AT^4 + BT^3 + CT^2 + DT + E$.

Age Range (in Ka)	CF	A	B	C	D	E
0–18	0	−0.005026	0.137217	−0.447917	0	0
18–28	0	0	0.1895	−12.993	283.554	−1871.9
28–35	0	−0.0336253	4.11303	−187.787	1796.45	−28655.8
35–45	30	0	0.0745	−2.1525	15.938	32.81
45–53	30	−0.0227332	1.60311	−41.513	471.838	−1957.77
53–58	50	0	0.906	−14.784	64.242	−6.63
58–74	50	−0.00278963	0.168136	−3.48014	30.763	−73.04
74–85	70	0	0.0812547	−2.24076	14.026	34.05
85–93	70	−0.0231131	1.55582	−37.982	402.962	−1564.75
93–107.5	90	0	0.0478658	−1.41498	7.197	62.35
107.5–116.5	90	−0.0159186	1.26071	−36.4263	459.548	−2138.6
116.5–122.5	110	0	0.731019	−20.6965	176.398	−402.92
122.5–130	110	−0.027282	1.67703	−37.8731	375.51	−1389.59
130–142	110	0	0.0282407	−2.13406	51.472	−379.26

In each interval, the age t (in ka) is modified by a correction factor (CF): $t - CF = T$.

and a mathematical expression to estimate sea level back to about 125 ka is shown in Table 6.1.

Uplifted coral reefs are not the only features used to infer ancient sea levels. The slow, steady buildup of sediment on the deep ocean floor and ice atop the Greenland and Antarctic ice caps record variations in naturally occurring isotopes of oxygen and carbon that are approximately synchronous with the coral-reef eustatic signal. Meteoric water (rainfall and snowfall) is depleted in the heavy isotope of oxygen (^{18}O) relative to ocean water; therefore the great accumulations of continental ice sheets that lowered sea level during the Pleistocene concentrated ^{18}O in the world's oceans [24]. A different sea-level driving mechanism operates for carbon. The light isotope of carbon (^{12}C) is concentrated in plants and other organic matter compared with its heavy stable isotope (^{13}C). Low sea levels expose the shallow continental shelves to erosion, flushing large amounts of light carbon into the oceans [24]. In fact, these oxygen and carbon isotope records provide a much more detailed history than do coral-reef sequences, suggesting a continuous record of sea-level changes through time. But the isotope histories are necessarily *indirect* measures of sea level. For example, the oxygen isotope ratio is also affected by the average temperature of the oceans as well as the latitude in which most precipitation falls; it is very difficult to translate a definite isotopic value into a particular sea-level elevation [25]. Uplifted coastal features are much more direct indicators of ancient sea level.

LONG-TERM UPLIFT

Long-term vertical movement of the land is the sum of all coseismic and inter-seismic movement (deformation during and between earthquakes, respectively) averaged over a long period of time. Under favorable conditions, long-term rates of uplift can be determined on clastic, carbonate, or erosional coasts. Favorable conditions include coastal landforms that are (1) durable and long-lived, and (2) capable of being dated.

As a general rule, clastic coasts are the least likely to meet the two criteria above. Beach ridges are loose and unconsolidated when they form, and any datable material originally incorporated into the deposits decays rapidly in this porous and unprotected setting. A good example of the limitations of and possibilities for using clastic coastal features in studying long-term uplift comes from the Atlantic coastal plain of the United States. The coastal plain has been emerging slowly from the Atlantic Ocean through the Pliocene and Pleistocene, leaving a path of old beach and dune ridges, escarpments, and marine-terrace remnants behind the retreating coast [26]. The shoreline features can be traced over long distances and, in places, can be correlated with the underlying marine stratigraphy to provide a rough chronology of their emergence. The slow but steady emergence of the Atlantic coastal plain is understood to be a result of erosion of the North American continent, deposition on the Atlantic continental shelf, and isostatic response of the Earth's crust [27].

In general, though, carbonate and erosional coasts tend to preserve a clearer picture of long-term uplift. The coral-reef sequences of Barbados and New Guinea are classic examples of emergent coastlines, reflecting the combined effects of long-term uplift and Pleistocene sea-level fluctuations. Continuous sequences of erosional marine terraces, such those along the Pacific Coast of North America, reflect the same processes at work.

Where such sequences of uplifted shorelines are used to infer former sea levels, a rate of uplift must be assumed; where the shorelines are used to infer the uplift rate, global sea-level history must be assumed (e.g., Figure 6.13). Uplift rate can be calculated using Equation 6.1, or determined graphically as shown in Figure 6.14. The graphical solution combines sea-level history plotted against time, the ages of preserved shoreline features, and the elevations at which those features currently are found. For example, a terrace formed at 85 ka when global sea level was 15 m below its current level (point A on Figure 6.14) is now found 88 m above modern sea level (point C at elevation B on Figure 6.14). The line connecting points A and B is the **uplift path** for that terrace. The uplift paths for all the terraces preserved in the same sequence should be parallel if the uplift rate in that location was uniform through time. The average slope of the uplift paths is the uplift rate [28].

Where many terraces can be identified in a single sequence, and where their elevations can be precisely determined, the uplift rate can be determined without dating every terrace in the sequence. Using the principles of the graphical correlation method illustrated in Figure 6.14, the **relative spacing** of the terraces is used to assign ages to each terrace in the sequence and an uplift rate to the site. Assumptions necessary for this method are that [28]

- heights of shoreline angles of terraces are accurately measured
- the sea-level model (ages and heights of highstands) used is correct and appropriate to the site
- the uplift rate at the site has been uniform though time

As a general rule, the higher an area's uplift rate, the better the resolution of terraces using the relative-spacing method. Each terrace in the sequence is ascribed

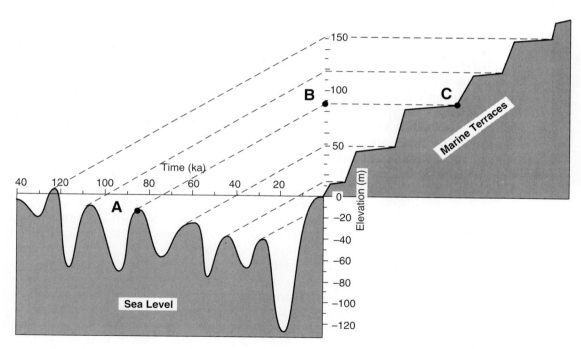

FIGURE 6.14
Graphical determination of the rate of uplift for a sequence of marine terraces (the terraces are shown on the upper right of the figure). The modern terraces (elevations at time = 0) are correlated to sea-level highstands during the Late Pleistocene. The line connecting the modern elevation of each terrace to the time and position at which it formed is its uplift path, the slope of which is the rate of uplift.

to a sea-level highstand appropriate to its elevation. If the result is not consistent with uniform uplift through time, then a different set of ages must be assigned until the result is consistent [28].

In most areas that have been studied, the assumption of a uniform uplift rate over periods of hundreds of thousands of years has proved satisfactory [29], but this is not always the case. Study of three marine-terrace sequences at widely separated sites along the Alpine fault in New Zealand [29] revealed a consistent history of uplift (Figure 6.15). Prior to about 140 ka, uplift was uniform (1.2 m/k.y.). However, between 140 and 135 ka, the uplift rate doubled to its present value of about 3 m/k.y. This change is interpreted to reflect accelerated convergence between the Australian and Pacific plates [30].

Long-term rates of uplift worldwide, measured over intervals of tens of thousands to hundreds of thousands of years, are greatest at convergent plate boundaries where sustained uplift rates of 6 m/k.y. to 10 m/k.y. have been calculated [31]. Transform boundaries show low uplift, generally less than 0.3 m/k.y. Intraplate and divergent boundary coasts show no movement to very low rates of uplift associated with erosion and isostatic rebound (see Chapter 9). The fastest uplift rates are ephemeral events related to temporary disequilibria in the Earth's crust; for example, **postglacial rebound** (uplift due to melting ice caps and the resulting isostatic compensation) and **volcanic tumescence** (uplift caused by magma-chamber inflation beneath a volcano). Whereas volcanic tumescence is a short-lived phenomenon of weeks to months, glacial rebound has occurred during the past 18 k.y., lifting Hudson Bay by over 250 m and is still lifting Scandinavia at a rate of up to 1 m per century (Figure 6.16) [32]. In fact, with the current rate of rebound along the shores of Lake Michigan, drainage from the lake will jump naturally from the St. Lawrence River to the Mississippi River within the next 3000 years [4].

FIGURE 6.15
Plot of uplift versus time for a sequence of marine terraces at Fiordland, New Zealand. Terraces formed prior to 135–140 ka indicate uniform uplift at a rate of 1.2 m/k.y., and terraces formed later indicate a rate two to three times more rapid.
(After Bishop, 1991 [30])

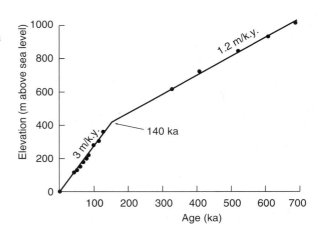

FIGURE 6.16
Modern rates of uplift (in
mm/yr) around the Baltic
Sea, caused by postglacial
unloading and isostatic com-
pensation.
(After Flint, 1971. *Glacial and
Quaternary Geology.* John Wiley &
Sons: New York)

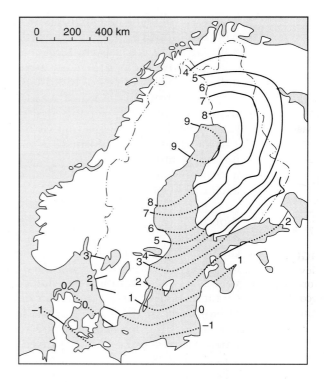

DEFORMATION OF MARINE TERRACES

In addition to the information that marine terraces and other coastal landforms
provide regarding uplift, either coseismic or long-term, these landforms also can
reveal patterns and sometimes rates of deformation of the surface. Most shoreline
features make good time-lines, having emerged from the maelstrom of the coastal
zone in an instant of geologic time. Furthermore, the shoreline angle of a marine
terrace and the crest of a beach ridge are excellent benchmarks of when sea level
formed. Together, these two facts make coastal landforms very sensitive indicators
of deformation, including both faulting and warping.

Faults are as likely to cut the surface of a marine terrace as they are to
cut any other landform. Unlike most of the Earth's surface, however, it is
sometimes possible to assign an age to marine terraces. If a fault cuts the surface of a given terrace, then the most recent movement on the fault is known
to *postdate* the age of the terrace [33]. The amount of vertical motion on the
fault is best estimated by measuring the displacement on the wave-cut platform, which is less subject to erosion than the surface of the terrace. On the
other hand, when the trace of a fault crosses a dated terrace but offsets neither
the surface nor the platform, then regardless of any amount of displacement

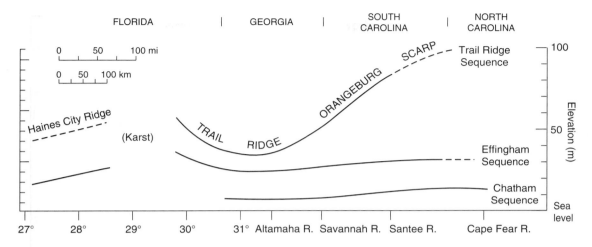

FIGURE 6.17
Elevations of strandlines along the Atlantic coastal plain south of the Cape Fear River,
North Carolina. Strandlines reflect ongoing uplift in the Cape Fear area. The magnitude of
recorded uplift increases with increasing age of the landforms.
(After Winker and Howard, 1977 [26])

on the underlying bedrock, the latest movement on the fault must *predate* the
age of the terrace.

Uplifted coastal landforms often can be traced or correlated over long distances. Variations in elevation of such a landform indicate variations in the rate of
uplift. Along the Atlantic Coast of the United States, the same, traceable Pleistocene strandline ranges in altitude from less than 50 m to about 100 m (Figure
6.17) in a systematic pattern reflecting a wave of postglacial rebound that has
migrated northward during the past 18 k.y. [26]. On the Osa Peninsula of Costa
Rica, Late Pleistocene beach ridges and nearshore marine deposits are uplifted by
amounts that decrease with distance away from the Middle America Trench (Figure 6.18), and merge with active subsidence in the gulf behind the peninsula [34].
Such landward tilting, across zones ranging in width from 30 to 50 km, seems to
be a characteristic process in the outer arcs of many subduction zones [15].

DATING COASTAL LANDFORMS

Like all studies in active tectonics, work on coastal landforms requires absolute
dates in order to calculate the rates of uplift or deformation. Fortunately, coastal
landforms sometimes incorporate material from the ocean, such as shells and
coral, that can be dated by one or more of several possible methods. Depending on
the nature of the material incorporated into a landform and its age, one or more of

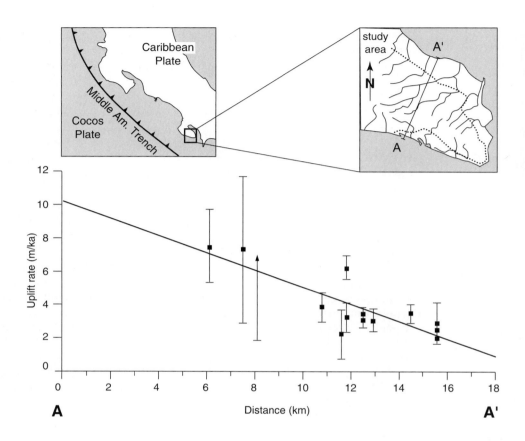

FIGURE 6.18
Uplift rate versus distance away from the Middle America Trench, Osa Peninsula, Costa Rica. The uplift rate declines to the northeast, becomes negative (subsidence) in the gulf, and probably represents underthrusting along the outer arc of the subduction zone.

the following dating techniques may be applicable: radiocarbon, uranium-series method, amino acid racemization, and electron spin resonance (see also Chapter 2, Pleistocene and Holocene Chronology).

The most straightforward coastal landform for dating is an uplifted coral reef. Unaltered coral as old as 350 ka can be dated using the uranium-series method. The ability to date uplifted marine terraces and beach ridges largely depends on the likelihood that carbonate material has been incorporated in deposits contemporaneous with the landform. For example, the marine veneer that usually overlies marine terraces may contain solitary corals that can also be dated using the uranium-series method [35]. Shells generally are a more complex material to date, but several relative and absolute methods have been developed to assign ages to shelly material. Among some assemblages, the variety of organisms preserved may contain a few species that indicate conditions either warmer or colder than the present, and thereby may suggest a landform age [14]. Finally, systematic processes of erosion may permit morphological dating of coastal landforms by comparing the degree of degradation (see discussion of Morphological Dating in Chapter 8). At the least, this last method is a useful tool for correlating from dated coastal landforms to undated equivalents in the same region [30].

It also should be noted that most tectonic analyses of shoreline landforms can be used for the shores of lakes as well. The Great Basin province of the United States is characterized both by repeated formation of Pleistocene lakes and by ongoing tectonic activity. Although lake formation and lake levels are primarily functions of climate, lake shorelines have been used as time lines to date fault movement [e.g., 36] and as horizontal benchmarks to determine deformation [e.g., 37]. Many of the techniques used to date coastal features, such as the uranium-series method and amino-acid racemization, can be used in lakes as well, although samples from isolated lakes may have different baseline conditions than in the much larger reservoir of the world's oceans. In addition, techniques are available for dating lake levels that do not exist for the oceans. For example, pack-rat middens (the nest and waste at the bottom of their dwellings) provide minimum ages for the extent of lakes in the Great Basin—because rats do not live under water, the oldest material accumulated in their nests must postdate the time that a lake retreated and exposed the site to air [38]. Other methods used to estimate ages of different lake levels include measuring the development of rock varnish, soils, and the accumulation of cosmogenic isotopes.

COASTAL TECTONICS AND TIME-SCALE

This chapter has discussed a number of geomorphic features of coasts that are formed by or are useful in studying active tectonics. The particular types of landforms preserved and the types of information they can yield are functions of the time-scale over which the formative processes have operated. Figure 6.19 summarizes this information.

Processes that operate over geologically short periods of time—for example, coseismic pulses of uplift, or catastrophic storms—leave their mark on the coastal area as beach ridges, wave-cut notches, or zones of uplifted intertidal organisms. Larger and more regional features, such as marine- and coral-terrace sequences, are the results of many smaller incremental events repeated over time, or the results of processes that operate over time spans of several thousand to tens of thousands of years, such as glacial-interglacial sea-level change. Finally, the end-product of sustained tectonic and geomorphic activity, the sum of both short-term and medium-term processes repeated through geologic time, is the construction of whole physiographic regions, from mountain chains to rift valleys. In time, individual landforms that were the record of short time-scale

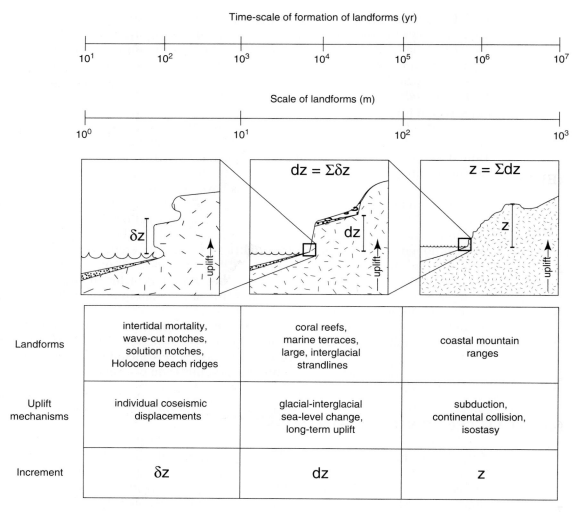

FIGURE 6.19
Variety and scale of coastal landforms formed by tectonic activity at different time-scales.
Processes that operate at longer scales of time are generally the sum of processes that
operate at short time-scales.

events are obliterated by erosion and are preserved only as anonymous incre-
ments in the construction of the topography of the Earth.

SUMMARY

At the boundary between land and water, coastlines consist of a distinct group of
landforms. The particular landforms that occur depend on the tectonic setting, the

sediment supply, erosional processes, biological processes, and the history of sea-level changes. Through time, coastal landforms may be deformed by tectonic activity. They can be uplifted, downdropped, faulted, or warped. Coastal landforms are particularly useful tools for studying tectonics for two reasons: (1) sea level acts as an absolute frame of reference, and (2) it is often possible to determine the age of coastal features. In addition, coastal landforms illustrate very clearly the role of time-scale in landscape evolution— processes that act over short periods of time blur together over longer and longer time intervals and create more regional elements of the landscape.

REFERENCES CITED

1. Carver, G. A., A.S. Jayko, D.W. Valentine, and W.H. Li, 1994. Coastal uplift associated with the 1992 Cape Mendocino earthquake, northern California. *Geology,* 22: 195–198.
2. Fischer, R., 1980. Recent tectonic movements of the Costa Rican Pacific coast. *Tectonophysics,* 70: T25–T33.
3. Norris, R.M., 1990. Sea cliff erosion: a major dilemma. *California Geology,* 43: 171–177.
4. Bloom, A.L., 1991. *Geomorphology: A Systematic Analysis of Late Cenozoic Landforms.* Prentice Hall, Englewood Cliffs, NJ.
5. Stoddart, D.R., 1969. Ecology and morphology of recent coral reefs. *Cambridge Philosophical Society Biological Review,* 44: 433–498.
6. Bloom, A. L., W.S. Broecker, J.M.A. Chappell, R.K. Matthews, and K.J. Mesolella, 1974. Quaternary sea level fluctuations on a tectonic coast: New $^{230}Th/^{234}U$ dates from the Huon Peninsula, New Guinea. *Quaternary Research,* 4: 185–205.
7. Bender, M.L., R.G. Fairbanks, F.W. Taylor, R.K. Matthews, J.G. Goddard, and W.S. Broecker, 1979. Uranium-series dating of the Pleistocene reef tracts of Barbados, West Indies. *Geological Society of America Bulletin,* 90: 577–594.
8. Eckel, E.B., 1970. The Alaska earthquake, March 27, 1964—lessons and conclusions. *U.S. Geological Survey Professional Paper* 546.
9. Emery, K.O., and D.G. Aubrey, 1991. *Sea Levels, Land Levels, and Tide Gauges.* Springer-Verlag: New York.
10. Plafker, G., 1965. Tectonic deformation associated with the 1964 Alaska earthquake. *Science,* 148: 1675–1687.
11. Bodin, P., and T. Klinger, 1986. Coastal uplift and mortality of intertidal organisms caused by the September 1985 Mexico earthquakes. *Science,* 233: 1071–1073.
12. Castilla, J.C., 1988. Earthquake-caused coastal uplift and its effects on rocky intertidal kelp communities. *Science,* 242: 440–443.
13. Pandolfi, J.M., M.R.B. Best, and S.P. Murray, 1994. Coseismic event of May 15, 1992, Huon Peninsula, Papua New Guinea: comparison with Quaternary tectonic history. *Geology,* 22: 239–242.
14. Lajoie, K.R., D.J. Ponti, C.L. Powell II, S.A. Mathieson, and A.M. Sarna-Wojcicki, 1991. Emergent marine strandlines and associated sediments, coastal California: a record of Quaternary sea-level fluctuations, vertical tectonic movements, climatic changes, and coastal processes. In R.B. Morrison (ed.), *Quaternary Nonglacial Geology: Conterminous U.S., DNAG vol. K-2.* Geological Society of America: Boulder, CO.
15. Ota, Y., 1986. Marine terraces as reference surfaces in late Quaternary tectonics studies: examples from the Pacific Rim. In W.I. Reilly and B.E. Harford (eds.), Recent crustal movements of the Pacific region. *Bulletin of the Royal Society of New Zealand,* 24: 357–375.
16. Atwater, B.F., and D.K. Yamaguchi, 1991. Sudden, probably coseismic submergence of Holocene trees and grass in coastal Washington State. *Geology,* 19: 706–709.
17. Clague, J.J., and P.T. Bobrowsky, 1994. Evidence for a large earthquake and tsunami

100–400 years ago on western Vancouver Island, British Columbia. *Quaternary Research,* 41: 176–184.

18. Atwater, B.F., M. Stuiver, and D.K. Yamaguchi, 1991. Radiocarbon test of earthquake magnitude at the Cascadia subduction zone. *Nature,* 353: 156–158.

19. Clarke, S.H. Jr., and G.A. Carver, 1992. Late Holocene tectonics and paleoseismicity, Southern Cascadia subduction zone. *Science,* 255: 188–192.

20. Darienzo, M.E., and C.D. Peterson, 1990. Episodic tectonic subsidence of Late Holocene salt marshes, northern Oregon Central Cascadia margin. *Tectonics,* 9: 1–22.

21. Atwater, B.F., 1992. Geologic evidence for earthquakes during the past 2000 years along the Copalis River, southern coastal Washington. *Journal of Geophysical Research,* 97: 1901–1919.

22. West, D.O., and D.R. McCrumb, 1988. Coastline uplift in Oregon and Washington and the nature of Cascadia subduction-zone tectonics. *Geology,* 16: 169–172.

23. Selby, M.J., 1985. *Earth's Changing Surface: An Introduction to Geomorphology.* Clarendon Press: Oxford.

24. Broecker, W.S., 1982. Glacial to interglacial changes in ocean chemistry. *Progress in Oceanography,* 11: 151–197.

25. Pinter, N., and T.W. Gardner, 1989. Construction of a polynomial model of sea level: estimating paleo-sea levels continuously through time. *Geology,* 17: 295–298.

26. Winker, C.D., and J.D. Howard, 1977. Correlation of tectonically deformed shorelines on the southern Atlantic Coastal Plain. *Geology,* 5: 123–127.

27. Colquhoun, D.J., G.H. Johnson, P.C. Peebles, P.F. Huddlestun, and T. Scott, 1991. Quaternary geology of the Atlantic Coastal Plain. In R.B. Morrison (ed.), *Quaternary Nonglacial Geology: Conterminous U.S., DNAG vol. K-2.* Geological Society of America: Boulder, CO.

28. Bull, W.B., 1985. Correlation of flights of global marine terraces. In M. Morisawa and J.T. Hack (eds.), *Tectonic Geomorphology. The Binghamton Symposia in Geomorphology: Inter-*

national Series, No. 15. Allen & Unruh: Boston.

29. Bull, W.B., and A.F. Cooper, 1986. Uplifted marine terraces along the Alpine fault, New Zealand. *Science,* 234: 1225–1228.

30. Bishop, D.G., 1991. High-level marine terraces in western and southern New Zealand: indicators of the tectonic tempo of an active continental margin. *Special Publications of the International Association of Sedimentology,* 12: 69–78.

31. Lajoie, K.R., 1986. Coastal tectonics. In *Active Tectonics.* National Academy Press: Washington, D.C.

32. Easterbrook, D.J., 1993. *Surface Processes and Landforms.* Macmillan Publishing Company: New York.

33. Keller, E.A., 1988. Estimating timing of fault activity on uplifted wave-cut platforms. *Bulletin of the Association of Engineering Geologists,* 25: 505–507.

34. Gardner, T.W., D. Verdonck, N. Pinter, R.L Slingerland, K.P. Furlong, T.F. Bullard, and S.G. Wells, 1992. Quaternary uplift astride the aseismic Cocos Ridge, Pacific Coast, Costa Rica. *Geological Society of America Bulletin,* 104: 219–232.

35. Stein, M., G.J. Wasserburg, K.R. Lajoie, and J.H. Chen, 1991. U-series ages of solitary corals from the California coast by mass spectrometry. *Geochimica et Cosmochimica Acta,* 55: 3709–3722.

36. Hanks, T.C., and R.E. Wallace, 1985. Morphological analysis of the Lake Lahontan shoreline and beachfront fault scarps, Pershing County, Nevada. *Bulletin of the Seismological Society of America,* 75: 835–846.

37. Bills, B.G., and G.M. May, 1987. Lake Bonneville; constraints on lithospheric thickness and upper mantle viscosity from isostatic warping of Bonneville, Provo, and Gilbert Stage shorelines. *Journal of Geophysical Research,* 92: 11,493–11,508.

38. Benson, L.V., D.R. Currey, R.I. Dorn, K.R. Lajoie, C.G. Oviatt, S.W. Robinson, G.I. Smith, and S. Stine, 1990. Chronology of expansion and contraction of four Great Basin lake systems during the past 35 ka. *Palaeogeography, Palaeoclimatology, Palaeoecology,* 78: 241–286.

7

Active Folding and Earthquakes

INTRODUCTION

Folds in layered rock form some of the most beautiful geologic structures on Earth (Figure 7.1). Folds are found in a variety of shapes and sizes, from microscopic to regional warps hundreds of kilometers across. Folds are described by a few simple terms (Figure 7.2). Each fold consists of two **limbs,** and the limbs join at the **hinge** of the fold. A fold hinge is defined as the line that connects points of maximum curvature along a fold. In a fold consisting of several rock layers, the surface that connects the different hinge lines is known as the **axial surface.**

Folds are classified into several different categories, depending on the orientation of their limbs. In the simplest case, where young rock layers over-lie older layers, an **anticline** is a fold in which the limbs slope down away from the hinge, and a **syncline** is a fold in which limbs slope toward the hinge (Figure 7.2). Anticlines and synclines are the two most basic and most

FIGURE 7.1
Photograph of folded rocks.
(Photograph courtesy of A.G. Sylvester)

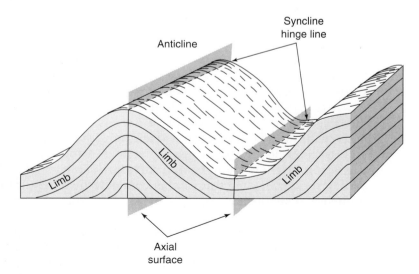

FIGURE 7.2
Idealized diagram showing the basic features of folds (an anticline and syncline).
(After Skinner and Porter, 1989. *The Dynamic Earth*. John Wiley & Sons: New York.)

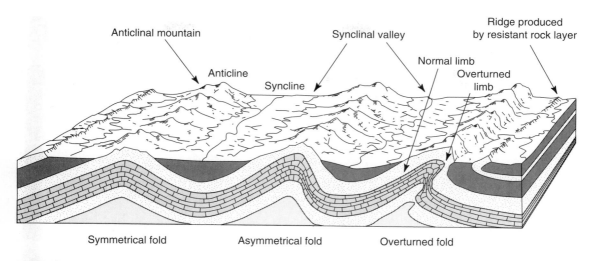

FIGURE 7.3
Block diagram illustrating several types of common folds, with possible surface expressions such as anticlinal mountains and synclinal valleys.
(After Lutgens and Tarbuck, 1992. *Essentials of Geology.* Macmillan: New York.)

common types of folds. Three additional complications also need to be considered; Figure 7.3 illustrates folds that are overturned, recumbent, and asymmetrical. In an **overturned fold,** the axial surface slopes such that both limbs dip in the same direction. In a **recumbent fold,** the axial surface is horizontal or nearly horizontal. Finally, in an **asymmeterical fold,** one of the limbs dips significantly more steeply than the other, and the axial surface is curved.

In Figures 7.2 and 7.3, all of the hinge lines for the folds are horizontal; this is seldom the case in nature. When the hinge line is inclined, we say that the fold **plunges.** A block diagram illustrating the concept of a plunging fold is shown in Figure 7.4A. Folds on geologic maps are represented by the hinge lines, and the **strike** (the compass direction of the intersection of the rock layer with a horizontal plane) and the **dip** (the maximum angle that the rock layer makes with the horizontal) of rock layers that crop out at the surface. A simple example of a geologic map is shown on Figure 7.4C. The hinge line and strike and dip symbols on the map show an anticline plunging to the northwest. The fold is slightly asymmetrical with the northeast limb dipping more steeply than the northwest limb.

Geologic structure can be evaluated by constructing geologic cross sections (Figure 7.4D) which aid in the interpretation of structures such as folds and faults. Subsurface control may come from well logs and geophysical data. Interpretation of cross sections of folds and faults is facilitated by using specific models to explain folding and to construct quantitative **balanced solutions** to cross sections. Such cross sections balance (maintain a constant total length and

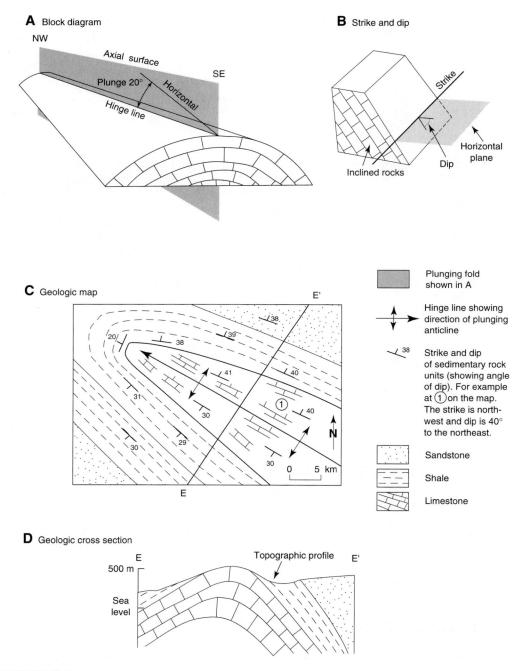

A Block diagram

B Strike and dip

C Geologic map

Plunging fold shown in A

Hinge line showing direction of plunging anticline

38 Strike and dip of sedimentary rock units (showing angle of dip). For example at ① on the map. The strike is northwest and dip is 40° to the northeast.

Sandstone

Shale

Limestone

D Geologic cross section

FIGURE 7.4
Idealized block diagrams of (A) a plunging fold, (B) strike and dip, (C) geologic map of plunging fold, and (D) geologic cross section.

thickness of rock units) before and after deformation [1, 2]. Other models have been developed to allow for deformation with changes in thickness of rock units [3]. Solutions to cross sections can be tested by **retrodeformation,** which removes the deformation, restoring the section to geologic conditions prior to folding and faulting. Another purpose of retrodeformation is to ensure that the "original" strata appear to make sense—to make sure that there are no voids in the undeformed section. For example, if one interprets the cross section in a particular geologic setting to be a tapered wedge of sediments that have been folded and faulted, then retrodeformation checks that the balanced section is viable. Comparison of deformed to retrodeformed conditions allows the amount of shortening due to folding and faulting to be estimated. If the ages of the deformed rocks are also known, then rates of tectonic processes can be estimated [4].

Construction of balanced cross sections is becoming a powerful tool for evaluating deformation and rates of folding and faulting. Having said that, it is important to understand that the use of balanced cross sections to produce models of subsurface geological structures (i.e., folds and faults) is controversial. There are serious issues related to this form of modeling, including:

- Availability of sufficient field, well log, and geophysical data to ensure accuracy of cross-section models.
- Strike-slip motion makes it difficult to produce viable cross sections.
- Methods of model constructions *require* the existence of relatively flat detachment faults at some depth. Therefore it is not surprising that the models *produce* these structures.
- The models may infer the existence of faults that, lacking surface deformation, may be difficult to evaluate (you cannot determine whether the faults are active).
- The models do not produce unique solutions (several solutions are possible that balance the cross section).
- Methods of deriving rates of tectonic processes are new and largely untested.

Nevertheless, models produced by construction of balanced cross sections are an exciting area of research that provides an important new dimension in understanding active tectonic processes.

FOLD-AND-THRUST BELTS

Folds are seldom found in isolation. Rather, folds are often found as belts of anticlines and synclines. Folds in an individual belt may plunge in one direction or another, and the hinge lines tend to be long relative to the width of individual folds. In addition, the fold belts themselves tend to be long and relatively narrow features. Folding of sedimentary layers is directly related to compressional stress and crustal shortening. Three principal causes of crustal shortening are:

- Plate tectonics at convergent plate boundaries
- Rotation of crustal blocks that cause one block to converge with another
- Strike-slip faulting where restraining bends and steps occur (see Chapter 2)

Shortening related to plate motion at a convergent plate boundary is illustrated in a cross section of the Cascadia Subduction Zone and associated fold-and-thrust belt of the Olympic Mountains (Figure 7.5). Farther south in Northern California, the fold-and-thrust belt bends and comes onshore near Eureka, where anticlines form hills and small mountains parallel to the faulting, while synclines are topographically low areas such as Humboldt Bay and Big Lagoon (review Figure 2.21 and accompanying discussion in Chapter 2).

Of particular importance to fold belts is the fact that the folds are commonly accompanied by reverse faulting. Many of these reverse faults are low angle (less than 45°) and are called **thrust faults,** so that the resulting structural domain is called a **fold-and-thrust belt.** Some of the thrust faults and high-angle reverse faults may break the surface, but many others remain hidden within the cores of anticlines and are termed **buried reverse faults.** Only recently has it been recognized that these buried active faults present a significant earthquake hazard—that large damaging earthquakes can occur on faults located entirely beneath or within folded rocks. Such faults may be buried at depths of several kilometers, and when they rupture during earthquakes, uplift and folding may occur at the surface [5]. The recognition of buried reverse faults and their accompanying earthquake hazard has resulted from the observation of several earthquakes in California and other areas that occurred on buried faults in anticlinal folds, without accompanying surface rupture. These events include the 1983 $M = 6.5$ Coalinga earthquake that caused $31 million in property damage, the 1987 $M = 5.9$ Whittier earthquake that claimed 8 lives and caused $358 million in damage, and the 1994 $M = 6.7$ Northridge (Los Angeles) earthquake that killed 61 people and caused more than $20 billion in damage.

Other large earthquakes in fold-and-thrust belts produce surface rupture as well as uplift and folding. For example, the 1980 $M = 7.3$ El Asnam, Algeria, earthquake that killed 3500 people was produced by 3 to 6 m of slip on a reverse fault several kilometers below the surface [5]. Ground rupture (as much as 2 m) reached the surface over a significant part of the fault, but the greatest deformation at the surface was anticlinal uplift of approximately 5 m. Figure 2.23 shows a schematic drawing of the fold as well as a topographic profile showing the uplift produced by the earthquake. The good correlation between the shape of the topographic uplift and the shape of the fold suggests that faulting during this earthquake and other events like it does produce folds [5].

The 1980 El Asnam earthquake produced a variety of near-surface deformations (Figure 7.6) [6]. The dominant deformation was upper-plate folding (folding

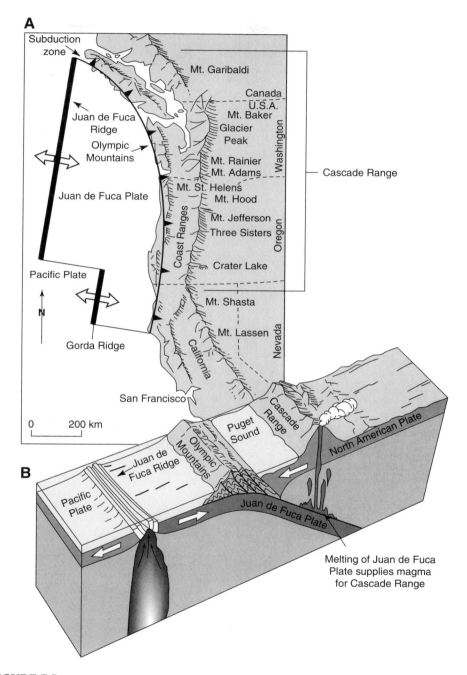

FIGURE 7.5

Cascadia subduction zone. The fold-and-thrust belt is just east offshore of the Puget Sound.
(From Coch and Ludman, 1991. *Physical Geology.* Macmillan: New York.)

above the fault plane; Figure 7.6A, B, D), but also included a variety of extensional features produced by a component of left-lateral shear, such as grabens (Figure 7.6D), tension fractures (Figure 7.6C), and elongated en echelon depressions (Figure 7.6A). Deformation also included lower-plate folding (folding beneath the

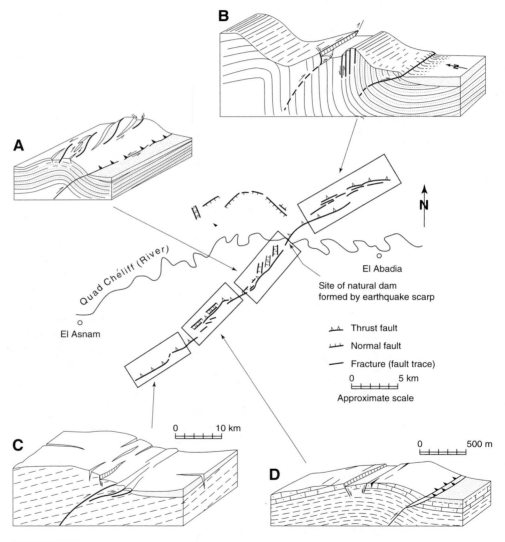

FIGURE 7.6
Generalized map of surface faulting and block diagrams showing characteristic deformation produced by the October 10, 1980, El Asnam, Algeria, earthquake in North Africa. (After Philip and Meghraoui, 1983 [6])

fault plane) and flexural-slip faulting (Figure 7.6B), which is discussed in a later section of this chapter. During the earthquake, the uplift and faulting blocked the Cheliff River, producing a lake upstream of the epicentral area. The river flowing into the lake deposited silt particles, producing a layer approximately 0.4 m thick. Small tectonic dams on rivers are not permanent features, and when the dam is breached, the silt deposits remain as evidence that a lake was there. Excavations in the El Asnam area revealed that there have been six such lakes in the past 6 k.y. Of course, additional earthquakes may have occurred but did not produce lakes. Smaller earthquakes might not cause sufficient uplift, and if a flood occurred very soon following the earthquake, then the dam might have been too short-lived for appreciable silt to have been deposited [5, 6].

As a result of recent earthquakes on buried faults, intensive study was begun on active folds to learn more about the relationship between buried faults and folds as well as the earthquake hazard that these structures present. For example, one of these belts occurs along the southern flank of the San Gabriel Mountains extending into the Los Angeles Basin. Buried faults present a significant earthquake hazard to the millions of people who live in the Los Angeles Basin [7, 8], and have produced several damaging earthquakes in the last 25 yr, including the 1994 $M = 6.7$ Northridge earthquake that caused over 60 deaths, damage to or destruction of thousands of buildings (Figure 7.7), and collapse of several freeway overpasses. The geometry of the Northridge earthquake is shown on Figure 7.8. The fault which caused the earthquake is inferred to be a reverse fault dipping 35° to 45° to the south. Rupture was initiated at a depth of about 18 km and quickly propagated upward (northward) up to a depth of about 4 km and laterally (mostly westward) about 20 km. The faulting and folding uplifted part of the Santa Susana Mountains (a few kilometers north of Northridge) about 38 cm and moved them 21 cm to the northwest [9]. Fault displacement near the focal depth of 18 km was about 2 m. Two other Southern California fold-and-thrust belts (Figure 7.9) are located south of Bakersfield, on the north flank of the Western Transverse Ranges, and near Ventura, on the southern flank of the ranges; these two belts are discussed in detail at the end of this chapter.

In order to understand relationships between faulting, folding, and fold-and-thrust belts, geologists have developed models for predicting the behavior of folding as the result of buried faults. The models are generally subdivided into:

- Fault-propagation folds
- Fault-bend folds

Fault-Propagation Folds

Fault-propagation folds are folds that form on the upper plate (**hanging wall**) of a thrust or reverse fault. Remember that a thrust fault is a reverse fault with a dip of less than 45°; the terms are loosely used and often are interchangeable (see

FIGURE 7.7
Damage to a parking structure at California State University, Northridge, produced by the
1994 Northridge earthquake.
(Photograph courtesy of F. Hopson)

Chapter 1). Fault-propagation folds result from deformation in front of a pro-
pogating fault [4]—that is, the tip of the fault is advancing the same way that a
newly formed crack advances across a car's windshield. Fault propagation folding at the tip of a thrust fault is idealized on Figure 7.10A. The propagating fault
tip (time 1 to time 3) diverges from a **décollement,** or **detachment fault,** that
runs along a weak stratigraphic horizon [4] and steps up through fault-bend
folds. Dashed lines are axial surfaces of the developing fold. Rocks below a
décollement are often relatively undeformed compared to rocks above the
décollement. As the fold develops, it becomes more asymmetric. Figure 7.10B
shows a fault-propagation anticline (the Meilin anticline) that is part of a fold-
and-thrust belt in western Taiwan, deforming Pleistocene strata. An important
observation associated with fault-propagation folding is that thrust faults and
folding are intimately related. Thrust faults commonly die out in folds that terminate a short distance from the end of the fault [4]. Furthermore, the rate of

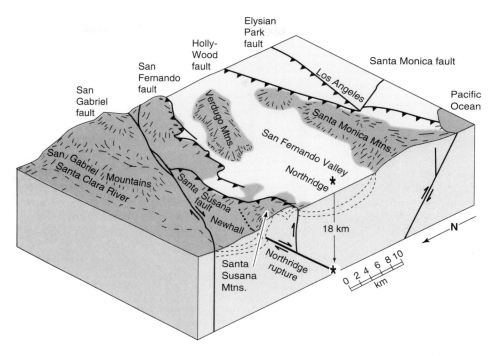

FIGURE 7.8
Block diagram of the San Fernando Valley showing the approximate location of the fault that generated the 1994 (*M* = 6.7) Northridge earthquake. Solid lines are strike-slip faults, and barbed lines are reverse or thrust faults, some of which, such as the Elysian Park and Hollywood, are buried.
(Courtesy of P. Williams and P. Holland, Lawrence Berkeley Laboratory. *Earth*, Sept. 94, p. 44)

lateral propagation of the fault and fold may be several times greater than the vertical slip rate of the fault [10].

Fault-Bend Folds

Fault-bend folds develop because faults are not perfectly planar surfaces. They may have significant curvature or even sharp bends ("ramps"). As slip occurs along irregular fault planes, fault-bend folding occurs. The two most commonly recognized structural environments for fault-bend folding are [4]:

- Steps or ramps along décollements (detachment faults)
- Flattening of normal faults with depth

Progressive development of an idealized fault-bend fold is shown on Figure 7.11. With progressive development of the fold, strata roll through the

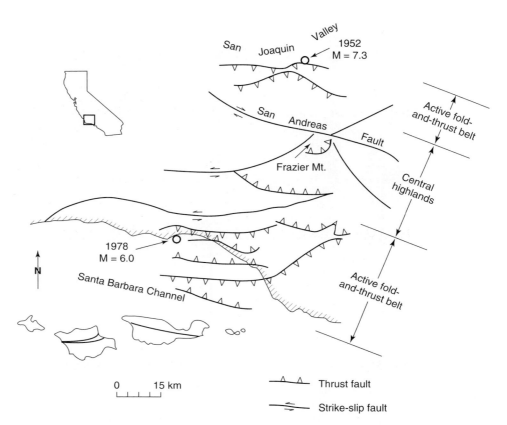

FIGURE 7.9
Map of fold-and-thrust belts in the Western Transverse Ranges.

fault bend, and axial surfaces become more widely spaced. The characteristic fault-bend anticline is nearly symmetric, but, as shown on Figure 7.11, the front limb may be steeper than the back limb. Notice that the folding is confined to the strata above the fault plane [4]. These flat-topped anticlines were described in 1934 by J. L. Rich [11] while interpreting low-angle thrust faulting of the Cumberland fault block in Virginia, Kentucky, and Tennessee. The geometry above the thrust fault for the Cumberland fault block is an anticline with a long, flat top, bordered by much shorter limbs. Structures like this are known as **monoclines**—structures characterized by a local steepening of dip in a sequence of layered rocks that otherwise dip gently.

Displacement on normal faults that bend (**listric faults**) and become flatter with depth are also associated with fault-bend folding on the down-thrown (**hanging wall**) block of the fault [4,12]. The term for such folding is **rollover** and the phenomenon is common in the Gulf Coast region and the

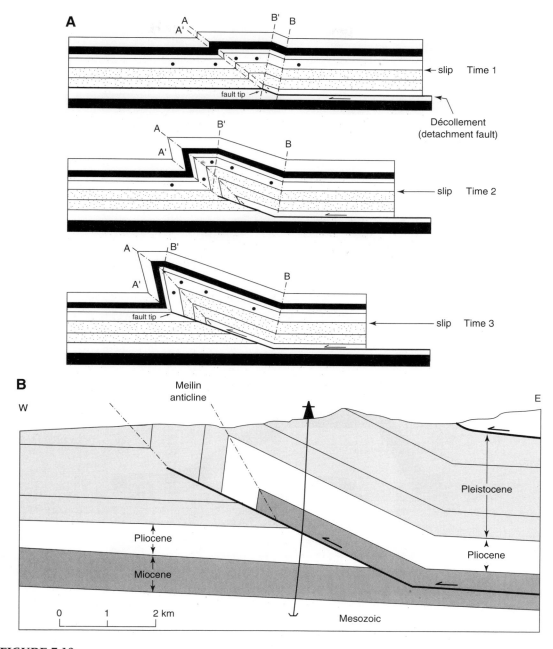

FIGURE 7.10
(A) Development of a fault-propagation fold as a result of slip along a décollement and a
steeper thrust fault. (B) Fault-propagation fold and the Meilin anticline in western Taiwan.
(From Suppe, 1985 [4])

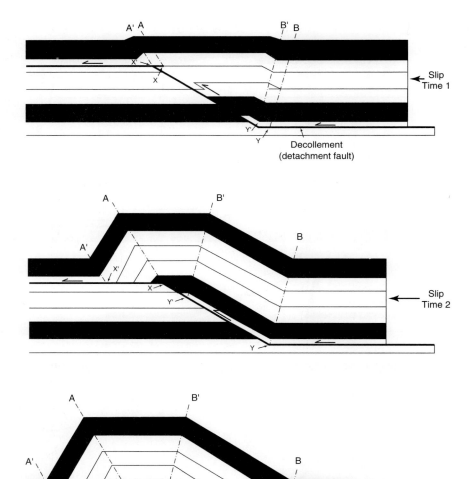

FIGURE 7.11
Progressive development (time 1 to time 3) of a fault-bend anticline.
(After Suppe, 1983 [2])

Colorado Plateau [12], as well as in other areas associated with extensional tectonics. The progressive development of a rollover fold along the Hurricane fault is shown on Figure 7.12. Folding had already started when extrusion of Quaternary age basalt (Q_{b1}, Figure 7.12A) covered the fault. Subsequently, the basalt was displaced by about 400 m and folded (Figure 7.12B), and a younger basalt (Q_{b2}) was extruded and is also displaced and folded (Figure 7.12C and D). Development of the rollover anticline is clearly related to collapse of the downthrown (hanging wall) block with time. Older strata are more strongly folded than the Quaternary basalt, indicating that faulting and folding occurred at the same time. Notice on Figure 7.12 that while the rollover anticline has beds dipping toward the fault, reversal of dip occurs at the fault due to drag. There is hot debate as to whether or not listric normal faults are capable of generating large earthquakes. In the Basin and Range province, listric faults are interpreted from seismic reflection profiles, but to date all large normal earthquakes have been on steeply dipping fault planes. If large, low-angle normal faults are capable of generating large earthquakes, then we are grossly underestimating the earthquake hazard of many regions [13]! Of course this is exactly what we did with respect to buried reverse faults in the Los Angeles area and other regions. Our historic record of large earthquakes on normal faults is too short to answer this important question.

Folds may also form over a buried normal fault, as idealized on Figure 7.13. Such folds are called **drape folds.** A good example of a young and active drape fold occurs on the east flank of the Sierra Nevada of southern Spain, near Granada. Normal faults are common along the range front, and several displace Quaternary deposits [14]. A drape fold is present near the toe of a prominent alluvial fan. Fan gravels generally dip 3° to 4°, but steepen to 10° to 12°, forming a scarp several tens of meters in height. This simple structure is a monocline, defined by steeply inclined strata bounded by more gently inclined strata. The pattern of deformation of the fan gravels suggests the presence of a buried normal fault below the alluvial fan. Figure 7.14 is a block diagram showing how displacement along a normal fault might change laterally to a monocline.

Fold-and-Thrust Belts: Selected Processes

Faulting with displacement along a décollement (detachment fault) is an important process in fold-and-thrust belts. For example, the décollement shown in Figure 7.15 displaces earth material in the opposite direction as the dip of the faults. Two asymmetric anticlines, formed by buried reverse faults, accommodate the shortening. The folds of the fault-propagation variety are said to **verge** to the south (Figure 7.15); vergence refers to the dominant direction of transport of material along both the décollement and the reverse faults that merge with the décollement. However, the situation is often complex, and there may be other reverse faults or back thrusts off the detachment surface that verge in the opposite direction. For example, the north-verging fault that generated the 1994 Northridge earthquake probably is a back thrust of the south-verging Santa Monica fault (see Figure 7.8). At convergent plate boundaries, the dominant

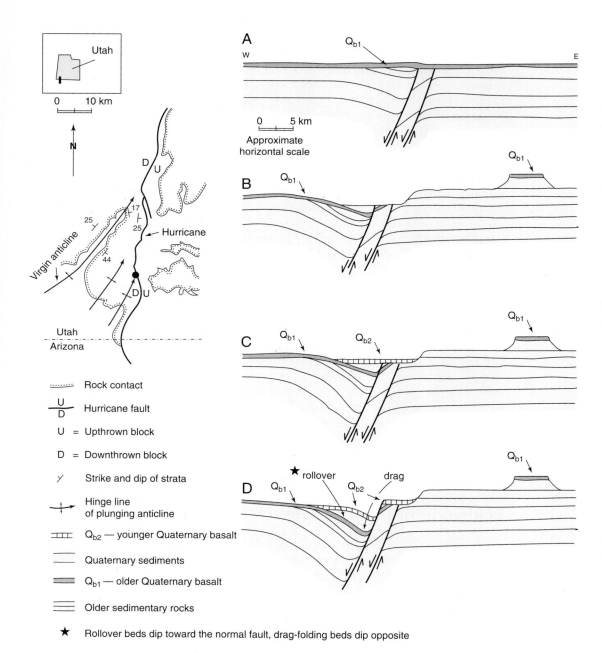

FIGURE 7.12
Progressive development of a rollover fold by collapse of the downthrown block (hanging wall) on the Hurricane normal fault, Utah and Arizona.
(After Hamblin, 1965 [12])

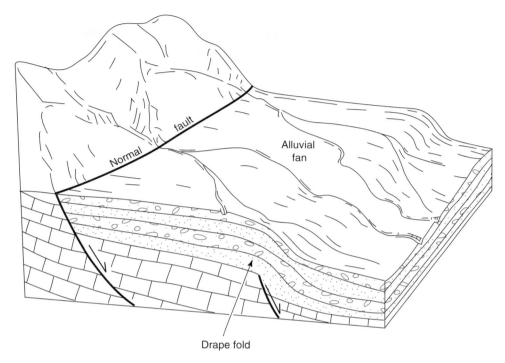

FIGURE 7.13
Idealized block diagram showing the development of a drape fold on alluvial fan gravels over a buried normal fault.

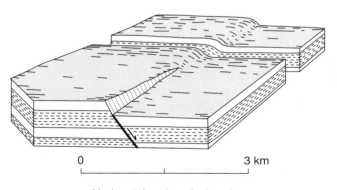

Horizontal and vertical scale

FIGURE 7.14
Block diagram showing how a normal fault could change laterally into a monocline, which is a simple fold where one limb dips steeply and the other limb is flat or gently dipping. (After Skinner and Porter, 1989. *The Dynamic Earth.* John Wiley & Sons: New York.)

FIGURE 7.15
Asymmetric fault propagation
folds over a décollement.

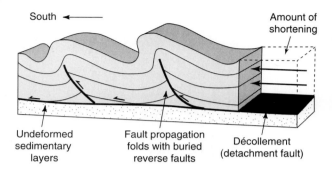

South ◄─────

Amount of shortening

Undeformed sedimentary layers

Fault propagation folds with buried reverse faults

Décollement (detachment fault)

direction of transport is toward adjacent basins. For example, in the continental collision forming the Himalayan Mountains, the dominant direction of transport is away from the mountain mass and toward the adjacent basins. Furthermore, the locus of tectonic activity usually migrates toward adjacent sedimentary basins. That is, as a mountain range forms, the locus of tectonic activity migrates away from the highlands of the range toward the adjacent flanks of the ranges. This migration widens the fold-and-thrust belt with time, and interior faults of the system may become inactive as the active tectonic processes are transferred to frontal fault systems [15, 16]. The pattern of deformation described above has been observed in fold-and-thrust belts in Taiwan (subduction zone) [17], India and Pakistan (continental collision forming the Himalaya) [16], Japan (subduction zone) [15], and California Transverse Ranges (shortening associated with San Andreas fault boundary) [8, 15, 16, 18, 19]. The observed pattern of thrust-fault migration is consistent with a mechanical model (Figure 7.16) proposed to explain the observed pattern of folding and faulting in Taiwan [17]. This model is based on earlier observations of onshore and offshore fold-and-thrust belts [20] that emphasize:

- Existence of a basal décollement (detachment fault) sloping toward the interior of the mountain belt below which there is little deformation
- The shape of the fold-and-thrust belt in cross section is a tapering wedge
- Existence of shortening in the tapering wedge

The mechanics operating in the model are analogous to those observed in the wedge of snow in front of a moving snowplow (Figure 7.16) [17]. The snow deforms until the wedge reaches a critical taper, when it locks up and slides over the basal décollement so that deformation is transferred to the front of the wedge where snow continues to accrete and deform [16, 17]. The model predicts that folds will migrate toward the edge of a fold-and-thrust belt, explaining why these folds and related faults are likely to be active and to present an earthquake hazard [16].

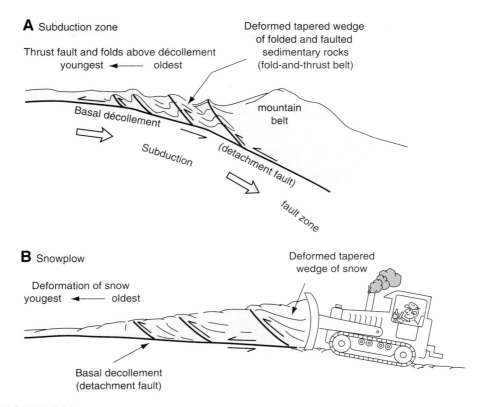

FIGURE 7.16
(A) Idealized diagram of a subduction zone and (B) analogy to a moving snowplow. With
time, new thrust faults form above the detachment near the narrower part of the tapered
wedge.
(After Davis et al., 1983 [17])

FLEXURAL-SLIP FOLDS

Flexural-slip folding occurs when the strain resulting from shortening is
concentrated directly along the surfaces that are being folded, such as the
bedding planes between sedimentary layers (Figure 7.17). It is important to
recognize that flexural-slip displacement along bedding planes reverses across
the hinge surface of a fold. Flexural-slip folding is analogous to bending a
telephone book so that displacement occurs along the pages. Because offset
does occur and the displacement surfaces are faults, they are termed **flexural-slip faults.**

The central Ventura Basin in Southern California has several good
examples of flexural-slip faulting and associated folding. A general tectonic
map of the region showing active and potentially active faults and folds is

FIGURE 7.17
Idealized diagram showing
flexural-slip folds.

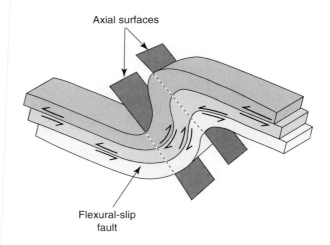

Axial surfaces

Flexural-slip
fault

shown on Figure 7.18A. The Devil's Gulch fault (Figure 7.18B) faults
Miocene shale over Ventura River terrace gravels (about 40 ka in age) with
several meters of displacement along a bedding plane in the shale [21]. Terraces of the Ventura River displaced by the flexural-slip faults are shown on
Figures 2.25 and 2.26.

One of the major thrust faults in the central Ventura Basin is the San
Cayetaño fault (Figure 7.18A), with stratigraphic separation of about 7.5 km
in the past 1 m.y. and therefore an average slip rate of about 7.5 mm/yr.
Study of Quaternary stratigraphy and tectonic geomorphology suggests a slip
rate of 1 mm/yr to 9 mm/yr [22]. Flexural slip is present at one location (Figure 7.18C). The style of deformation at that site is very similar to deformation
during the 1989 El Asnam event (see Figure 7.6B), suggesting that the flexural slip may be coseismic with large earthquakes on the San Cayetaño fault.

Flexural-slip faults, such as those in the central Ventura Basin, can
cause a ground-rupture hazard. It is debatable, however, whether flexural-
slip faults also present a seismic-shaking hazard. If the fold structures are
shallow, then these faults probably do not by themselves produce large earthquakes capable of strong seismic shaking, because the faults may not extend
to depths where large earthquakes are generated. On the other hand, the
folding is probably related to deeper, buried structures that do produce seismic shaking during earthquakes when the folding occurs. Thus, because the
flexural-slip faults themselves may not be capable of producing earthquakes
does not mean they are not associated with seismic shaking. Most likely, flexural-slip faulting accompanies folding during earthquakes (e.g., during the
$M = 7.3$ El Asnam event) and causes ground rupture that presents an additional hazard.

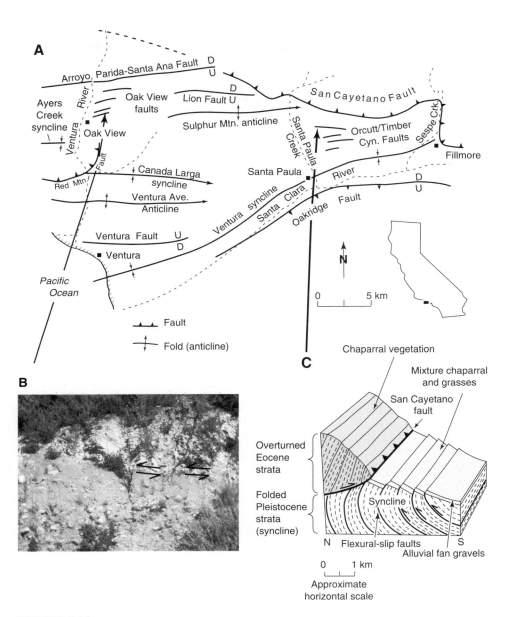

FIGURE 7.18
(A) Map of the central Ventura Basin, California, showing major geologic structures that
deform upper Pleistocene and/or Holocene materials. V = Ventura, OV = Oak View,
SP = Santa Paula, and F = Filmore. (B) Devil's Gulch fault, one of the Oak View flexural-slip
faults. Distance across the photograph is several meters. (C) Block diagram of flexural-slip and
thrust faulting associated with the San Cayetano fault and Ventura syncline. Compare with
similar deformation produced by the October 10, 1980, El Asnam earthquake (Figure 7.6B).
(Photograph by E.A. Keller)

FOLDING AND STRIKE-SLIP FAULTING

Where different segments of strike-slip faults overlap, shortening between the segments occurs if the step-over is in the opposite sense of the slip on the fault. For example, two strands of the southern San Andreas fault in the Indio Hills, not far from Palm Springs, California, overlap and form a left-step (Figure 7.19). That is, if you go to the end of either one of the fault segments in the central part of the map, you would have to turn left to get back to another strand of the fault. Within this area of the left-step in the right-lateral San Andreas system, convergence occurs (see Figure 2.12). The cross section in Figure 7.19 shows several folds extending into and through the area of overlap of the fault segments. The folded rocks are geologically young sediments with an age of about 1 Ma [23]. The convergence or shortening that produces the folds is directly related to the fault slip on the two segments.

Folds also can form as a result of **strain partitioning.** Oblique strain in the lower crust can be partitioned into nearly pure tangential and normal strain in the upper seismogenic crust (the upper 10 km to 15 km or so) [24]. For example, it has been observed that oblique convergence in subduction zones often results in partitioning into nearly pure thrust-faulting and folding as well as nearly pure strike-slip faulting. Similarly, along transform-fault plate boundaries, oblique strain is partitioned into (1) nearly pure strike-slip, and (2) shortening perpendicular to the strike-slip that results in fold-and-thrust belts roughly parallel to the strike-slip faults (Figure 7.20A). If the strain partitioning is a local phenomenon (Figure 7.20B) that occurs at shallow depths above where large earthquakes nucleate, then the subparallel faults and folds should be considered as a group when evaluating earthquake hazard. However, if the strain partitioning occurs regionally (Figure 7.20B), then individual thrust faults can probably be treated as independent seismic sources [24]. We state "can probably" because we really don't know for sure. The 1957 Gobi-Altay earthquake in China was associated with regional strain partitioning. A zone of about 250 km experienced dominant strike-slip displacement of 3–6 m while another 100-km-long zone about 30 km south experienced reverse fault displacements (vertical components 1–3 m). If a similar event in Southern California were to occur, it could cause simultaneous rupture of the San Andreas fault north of Los Angeles with one of the reverse faults in the Los Angeles Basin [25]. Certainly we know the reverse faults in the Los Angeles area do rupture independently of the San Andreas fault, presenting a serious earthquake hazard [26]. What we don't know for sure is what will happen when the San Andrea fault ruptures.

The above discussion argues for growth of folds during earthquakes. This is called **coseismic folding.** There is ample evidence that coseismic folding is the dominant process by which folds grow in fold-and-thrust belts of the world. On the other hand, where faults creep, uplift and folding may be a more continuous process, known as **aseismic folding** (that is, without recorded earthquakes). For example, resurveying of a level line across and perpendicular to the San Andreas fault at Durmid Hill, located adjacent to the northeast shore of the Salton Sea, suggests that the hill domed upward by 9 mm between September 1985 and

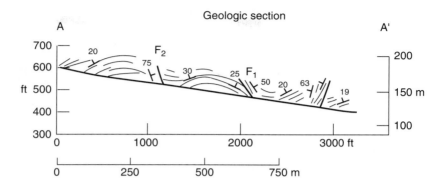

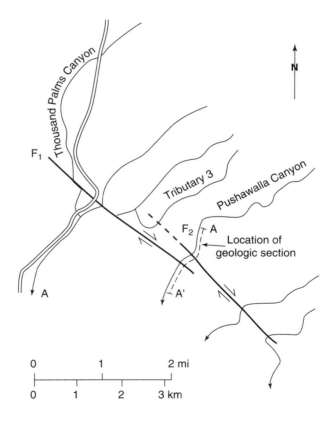

FIGURE 7.19

Sketch map and geologic section of the Pushawalla Canyon area. Folds are present in the area of the left-step of the San Andreas fault.

(From Keller et al., 1982 [23])

A

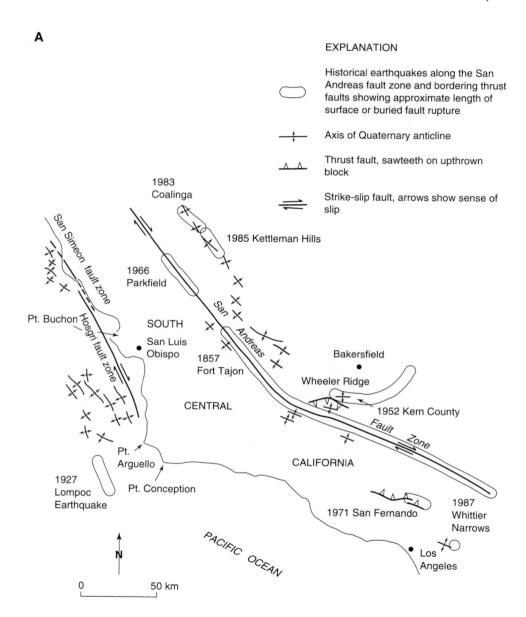

EXPLANATION

Historical earthquakes along the San Andreas fault zone and bordering thrust faults showing approximate length of surface or buried fault rupture

Axis of Quaternary anticline

Thrust fault, sawteeth on upthrown block

Strike-slip fault, arrows show sense of slip

1983 Coalinga

1985 Kettleman Hills

1966 Parkfield

San Simeon fault zone

Pt. Buchon

Hosgri fault zone

SOUTH

San Luis Obispo

1857 Fort Tajon

CENTRAL

San Andreas

Bakersfield

Wheeler Ridge

1952 Kern County

Fault Zone

CALIFORNIA

Pt. Arguello

Pt. Conception

1927 Lompoc Earthquake

1971 San Fernando

1987 Whittier Narrows

PACIFIC OCEAN

Los Angeles

N

0   50 km

FIGURE 7.20

(A) Generalized map of the San Andreas and San Simeon–Hosgri fault systems showing selected historical earthquakes and Quaternary anticlines. Note that the hinge lines are parallel to the strike-slip faults, suggesting that there is shortening perpendicular to the faults.

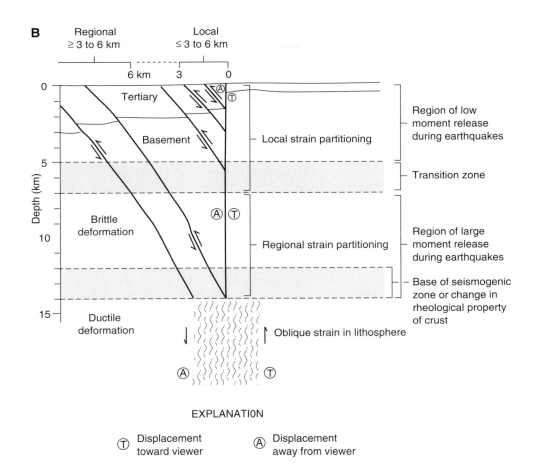

FIGURE 7.20 *(continued)*
(B) Idealized diagram of how oblique strain in the lithosphere may be partitioned at local or regional scales. The vertical fault is nearly pure strike-slip, whereas the dipping faults are nearly pure thrust (shortening perpendicular to the strike-slip). Earthquakes are generated in the brittle zone above the ductile zone. In the ductile zone, oblique strain occurs by processes such as slow flowage without brittle fracture.
(After Lettis and Hanson, 1991 [24])

December 1991 (about 1.5 mm/yr). The rate of aseismic strike-slip motion (creep) is between 2 mm/yr and 6 mm/yr. These rates are consistent with longer-term rates determined from stratigraphic studies, suggesting that coseismic deformation is not necessary to produce Durmid Hill. Local strain partitioning of oblique creep is sufficient to produce the 8.5 m of relief (elevation difference base to top) of the hill [27]. However, this certainly doesn't *preclude* coseismic folding at the site—only that at present creep rates, it is possible to produce the hill since mid-Pleistocene time (last 740 k.y.) without deformation from large earthquakes.

TECTONIC GEOMORPHOLOGY OF ACTIVE FOLDS

Active folding may occur well below the surface of the Earth, and if the uplift related to that folding is less than the rate of deposition of sediment, then it will not be observed at the surface. On the other hand, active folding may also cause direct deformation at the surface of the Earth that may be observed and measured. In order to evaluate the potential earthquake hazard associated with buried faults that produce folds, it is necessary to evaluate the evidence for active folding at the surface. If a fold stopped growing prior to the Pleistocene (last 1.65 m.y.), then the buried fault producing the fold is probably no longer active. Two case histories illustrate selected aspects of active folding:

- Ventura Avenue anticline, which includes a series of folded Late Pleistocene river terraces
- Wheeler Ridge anticline, which folds Late Pleistocene and Holocene alluvial fan deposits

Both of these folds are located in Southern California fold belts (see Figure 7.9). The Ventura Avenue anticline is on the south flank of the Western Transverse Ranges; the Wheeler Ridge anticline is on the north flank of the same mountain range, in the southern San Joaquin Valley.

CASE STUDY ══

VENTURA AVENUE ANTICLINE

The Ventura Avenue anticline is a fold located about 4 km north of Ventura, California (Figure 7.21). Subsurface evidence suggests that the fold is forming above a décollement in Miocene rocks at a depth of approximately 5 km (Figure 7.22) [16]. The anticline is the landwardmost fold of an active fold-and-thrust belt along the Pacific Coast. Rocks in the anticline are of Pliocene and Pleistocene age. Figure 7.23 shows the stratigraphy within the anticline, including the sequence of river terraces that cross the structure and are deformed by it (Figure 7.24). This Pleistocene stratigraphic section is one of the best-dated in the world. Table 7.1 shows rates of uplift associated with the Ventura Avenue anticline. The rates are based upon absolute ages for the river terraces.

Study of the Ventura Avenue anticline [28] suggests that rates of tectonic activity may vary in time and space as a result of the mechanics of folding. Folding started as recently as about 200 ka to 400 ka, and the rate of uplift atop the anticline during that time was approximately 7 to 14 mm/yr. The uplift rate today, however, is closer to 4 mm/yr. The shortening that causes this deformation is thought to be constant, at a rate of about 10 mm/yr. The

FIGURE 7.21
Photograph of the Ventura Avenue anticline.
(Courtesy of A.G. Sylvester)

FIGURE 7.22
Décollement below the Ventura Avenue anticline (VA).
(After Yeats, 1986 [16])

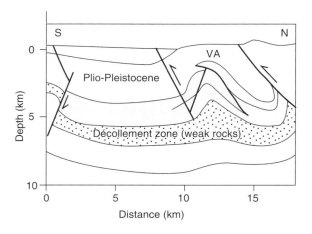

decrease in the uplift rate may be explained by the fact that as the rootless fold buckles, the mechanics of the process dictate that uplift decreases with time (Figure 7.25). Figure 7.25 reveals some surprising information concerning the mechanics of folding above a detachment—notice that 4% of the shortening

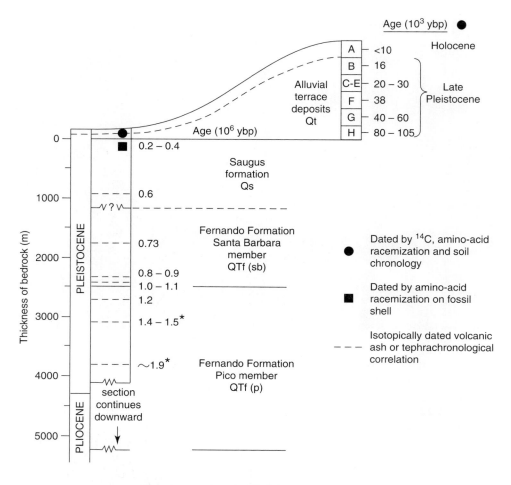

FIGURE 7.23
Pleistocene stratigraphy and alluvial (river) terraces at the Ventura Avenue anticline.
(Pleistocene stratigraphy, in part, from Lajoie et al., 1982. In *Neotectonics in Southern California.*
J.D. Cooper, compiler, Cordilleran Section, *Geological Society of America Field Trip Guidebook;* Rockwell
et al., 1988 [28])

produces approximately 36% of the uplift, and 8% of the shortening produces one-half of the total uplift! The rate of shortening is constant, but the uplift produced for each increment of shortening decreases with time, so that the uplift rate decreases.

The fact that a constant tectonic driving force can produce different uplift rates over time has important implications for interpretation of rates of active tectonics. It implies that when evaluating folds, the rate of uplift may be somewhat misleading, particularly if

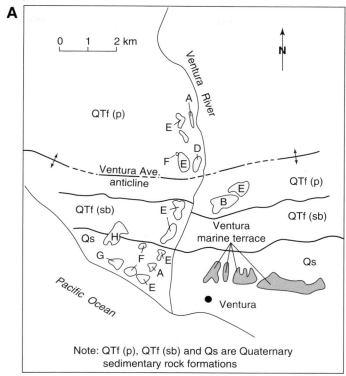

A

0   1   2 km

N

QTf (p)

Ventura River

A

E

D

F E

Ventura Ave.
anticline

QTf (p)

QTf (sb)

E

B

E

Ventura
marine terrace

QTf (sb)

Qs

H

G   F   E
      A

E

Pacific Ocean

Qs

● Ventura

Note: QTf (p), QTf (sb) and Qs are Quaternary
sedimentary rock formations

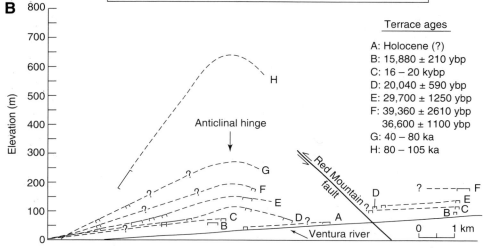

B

Elevation (m)

800
700
600
500
400
300
200
100
0

H

Anticlinal hinge

Red Mountain fault

G

F

E

D

C

B

?

?

?

?

?

D?

A

Ventura river

F

?

D

E

B   C

0   1 km

Terrace ages

A: Holocene (?)
B: 15,880 ± 210 ybp
C: 16 – 20 kybp
D: 20,040 ± 590 ybp
E: 29,700 ± 1250 ybp
F: 39,360 ± 2610 ybp
   36,600 ± 1100 ybp
G: 40 – 80 ka
H: 80 – 105 ka

FIGURE 7.24
(A) Map of Ventura River terraces over Ventura Avenue anticline and (B) terrace profiles
along the Ventura River.
(From Rockwell et al., 1988 [28])

TABLE 7.1
Uplift of river terraces over the Ventura Avenue anticline. Locations of terraces are shown on Figure 7.23.

River Terrace	Age*	Present Height Above Ventura River (m)**	Best Estimated Uplift Rate (mm/yr)***
B.	15,800 ± 210	30.5 ± 10	4.25 ± 0.7
D.	20,040 ± 590	85.3 ± 10	5.65 ± 0.65
E.	29,700 ± 1250	120 ± 10	4.50 ± 0.5
F.	38,000 ± 1900	175 ± 10	4.95 ± 0.5
H.	80,000 or 105,000	625 ± 100	7.10 ± 2.0

*Based on ^{14}C and amino-acid racemization chronology
**Projected to the hinge of the anticline
***Uplift rate determined using the best estimated depths of incision of the Ventura River

(Data from Rockwell et al., 1988 [28])

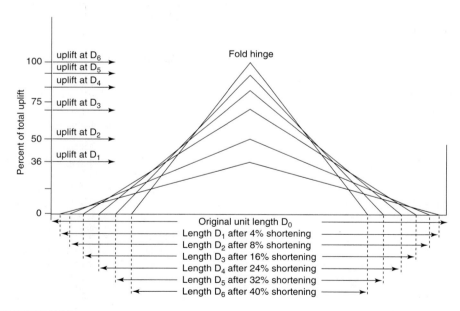

FIGURE 7.25
Simple model showing that, with folding above a décollement, a large amount of the uplift is produced early in the folding process. Maximum uplift occurs at D_6 with 40% shortening, as the fold locks up and deformation is transferred to another structure.
(After Rockwell et al., 1988 [28])

the limbs of the fold dip steeply, which would produce relatively low uplift rates. In contrast, if the fold is just beginning to form, uplift rates may be relatively high. You can verify the mechanics of folding that result in a fast initial rate of uplift followed by lower rates with a simple experiment. Use a thin, flexible straight-edge ruler with a length of about 30 cm and shorten the ruler (holding one end fixed) 1 cm at a time while recording the vertical deformation (uplift). Make a graph with uplift (in centimeters) on the vertical axis (x) and amount of shortening (in centimeters) on the horizontal axis (y).

In summary, the Pleistocene terraces of the Ventura River are clearly uplifted, tilted, and folded over the Ventura Avenue anticline. Rates of uplift and tilt have decreased since folding began, from ap-proximately 7 to 14 mm/yr to about 4 mm/yr [28]. The limbs of the Ventura Avenue anticline dip as steeply at about 45° on the south flank of the fold, and at this dip additional shortening produces little uplift. As the Ventura Avenue anticline begins to lock up, deformation may be transferred to the Ventura fault, a few kilometers to the south. At present, that fault is a buried feature, with little apparent displacement near the surface, but it is likely to become more active in the future, presenting a potentially serious threat to the Ventura area. An important additional point is that the variable rates of uplift on the Ventura Avenue anticline are for one particular type of rootless fold; uplift of a fault-propagation fold that is not rootless would probably be a different story.

CASE STUDY

WHEELER RIDGE ANTICLINE

The east-west–trending Wheeler Ridge anticline is located on the north flank of the San Emigdio Mountains in the southern San Joaquin Valley of California. Wheeler Ridge (Figure 7.26) is the northernmost topographic expression of the fold-and-thrust belt located on the northern flank of the Western Transverse Ranges [18]. The structural geometry of active folding in the Western Transverse Ranges and the southern San Joaquin Valley has been described in terms of both fault-bend folding and fault-propagation folding [18, 24, 29]. Several valleys have eroded through the fold. The prominent valley in the central part of Figure 7.26A is an abandoned stream valley. The stream in the valley was *defeated* by uplift of the fold approximately 30 ka to 60 ka, as determined by soil analysis and absolute and relative dating [10]. By defeated, we mean that the stream was able to cut its valley fast enough to keep up with the uplift of the fold for a while, but eventually the stream could no longer keep up with the uplift, or it was diverted around the low end of the fold to the east, and the valley was abandoned. The abandoned valley is a landform known as a **wind gap.** Farther to the east is a **water gap,** where the stream flowing from north of the ridge into the San Joaquin Valley is still eroding through the structure (Figure 7.26). In the central portion of the fold, the subsurface faulting that produced the structure is a thought to be a complex fault wedge [29]. Near the eastern end of the structure, the Wheeler Ridge thrust comes closer to the surface and the fold may be described as a fault-propagation feature (Figure 7.27).

Near the eastern end of Wheeler Ridge, a thrust fault with the expected geometry of

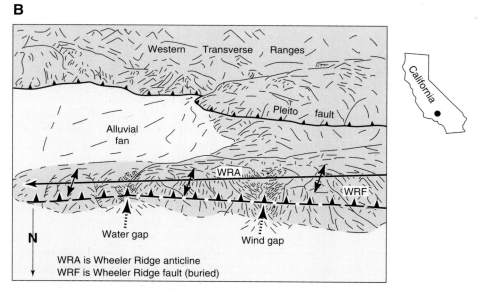

FIGURE 7.26

(A) Oblique aerial photograph of Wheeler Ridge. The view is to the south, and the fold is in the central portion of the photograph. Note that the fold is strongly asymmetric—the northern limb is more steeply inclined than the southern limb. (B) Sketch map of area shown in aerial photograph.

(Photograph by J. Shelton)

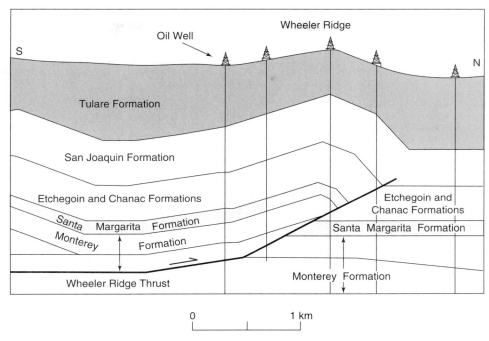

FIGURE 7.27
Cross section near the eastern end of Wheeler Ridge. To the west of this section the fault
geometry is more complex and is described as a fault wedge.
(After Medwedeff, 1984. Unpublished report)

the buried Wheeler Ridge fault is exposed in a
gravel pit. A sketch of the exposure (Figure
7.28) illustrates that near the surface, a soil
horizon is offset approximately 2.5 m. The
displacement increases dramatically with
depth, and at about 15 m, sedimentary units
are folded and dip as steeply as 45°. These
steep dips are consistent with deformation
associated with the Wheeler Ridge anticline.
This fault exposure illustrates an important
point—with active folding, the expression of
folding and fault displacement may increase
significantly with depth. Thus, information
obtained from shallow trenches excavated to
expose faults in anticlines may be very misleading when compared to displacement and
dip only a few meters below.

Evaluation of the deformed alluvial fan
deposits at Wheeler Ridge [29, 10] suggest
that the rate of uplift of the structure is
approximately 3.2 mm/yr and that the fold is
propagating to the east at a rate of about 29.0
mm/yr. The rates of uplift and propagation of
the anticline were determined through careful dating of several parts of the fold. Absolute
dating, soil chronology, and topographic data
revealed that alluvial fan segments folded
over the structure are lower and younger to
the east. Propagation of the fold to the east is
also revealed on aerial photographs; notice on
Figure 7.26A that the **drainage density**
(ratio of total length of streams to area) is
lower in the east. Drainage density tends to
increase with time and relief, so the decrease

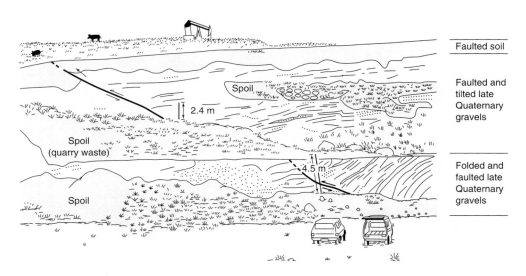

Faulted soil

2.4 m

Spoil

Faulted and
tilted late
Quaternary
gravels

Spoil
(quarry waste)

4.5 m

Folded and
faulted late
Quaternary
gravels

Spoil

FIGURE 7.28
Sketch map of exposure of presumed Wheeler Ridge thrust fault in gravel pit at east end
of Wheeler Ridge anticline.
(From Zepeda et al., 1991 [10])

in drainage density is additional evidence that
the eastern portions of Wheeler Ridge began
to fold more recently than the western portions [30].

By combining the evidence of Holocene
and latest Pleistocene folding and faulting
near the east end of Wheeler Ridge with subsurface information, the earthquake hazard
of this fold system may be evaluated. In the
Wheeler Ridge area at present, active tectonics is partitioned between the Pleito and
Wheeler Ridge faults. The Pleito fault, located
south of Wheeler Ridge, breaks the surface
and has an uplift rate of approximately 0.5
mm/yr [31]. The Pleito fault coincides with
the break-in-slope at the foot of the mountains surface south of Wheeler Ridge in Figure 7.26. The Wheeler Ridge fault is a buried
reverse fault with an uplift rate of approximately 3.2 mm/yr and is well defined by subsurface data [29]. The 1952 $M = 7.5$ Kern
County earthquake was located beneath

Wheeler Ridge on a deeply buried member
of the Pleito thrust fault system [29, 32]. In
the epicentral region, at Wheeler Ridge, the
earthquake produced approximately 1 m of
uplift. Assuming that $M = 6$ to $M = 7$ earthquakes in the Wheeler Ridge area are associated with approximately 1 m of uplift, then
the uplift rate of Wheeler Ridge suggests an
average recurrence of approximately 300
years for such events (1000 mm divided by
an uplift rate of about 3.2 mm/yr = 300 yr).
However, the earthquake history is probably
much more complex, and as a result an
average recurrence interval is probably not
very useful. There are several faults present
[32], and it is unlikely that each earthquake
produced a uniform 1 m of uplift. More
likely, both uplift and recurrence intervals
are variable. What we *can* say is that large,
damaging earthquakes can be expected to
occur in the future along this active fold-
and-thrust belt.

SUMMARY

Folds are found in a variety of shapes and sizes, and there is a generally-accepted terminology for describing them. Construction of geologic cross sections aids in the interpretation of subsurface geological structures such as folds and faults. Interpreting cross sections is facilitated by developing models that provide quantitative, geometrically balanced solutions. These models are generally subdivided into fault-propagation folds and fault-bend folds. Fault-propagation folds form on the upper plate of a thrust fault and result from deformation in front of a propagating fault surface. The two most common occurrences of fault-bend folding are bends (ramps) in detachment faults and flattening of normal faults with depth (listric normal faults). It is important to understand, however, that the use of balanced cross sections is controversial, and there are serious questions about the assumptions involved in this form of modeling.

Flexural-slip folding occurs when the strain resulting from shortening is concentrated directly along the surfaces that are being folded. This type of folding is analogous to bending a telephone book so that displacement occurs along the pages. Folding is often associated with strike-slip faulting. Where segments of strike-slip faults overlap or a segment bends, shortening results if the overlap or bend is in the opposite sense of the slip of the fault. Folds can also form as a result of strain partitioning. Along transform-fault plate boundaries, oblique strain is partitioned into nearly pure strike-slip parallel to the fault and shortening perpendicular to the fault.

Seldom is only a single fold present; rather, a series of folds often defines a fold-and-thrust belt. Some of the thrust faults and high-angle reverse faults may break the surface, but many others may be hidden within the cores of anticlines and are called buried reverse faults. Only recently has it been recognized that these buried active faults present a significant earthquake hazard. Buried reverse faults, concealed within folds, are difficult to evaluate. As a result, we study the folding above these buried faults. If the folding can be shown to be active, then we can infer that the concealed fault also is active. If we can calculate the rate of uplift of the folding, we can use that information to calculate the slip rate of the concealed fault.

REFERENCES CITED

1. Dahlstrom, C.D.A., 1969. Balanced cross sections. *Canadian Journal of Earth Sciences,* 6: 743–757.
2. Suppe, J., 1983. Geometry and kinematics of fault-bend folding. *American Journal of Science,* 283: 684–721.
3. Mitra, S., 1992. Balanced structural interpretations in fold-and-thrust belts. In S. Mitra and G.W. Fisher (eds.), *Structural Geology of Fold-and-Thrust Belts.* Johns Hopkins Press: Baltimore, MD.

4. Suppe, J., 1985. *Principles of Structural Geology.* Prentice Hall: Englewood Cliffs, NJ.

5. Stein, R.S., and R.S. Yeats, 1989. Hidden earthquakes. *Scientific American,* 260 (60): 48–57.

6. Philip, H., and M. Meghraoui, 1983. Structural analysis and interpretation of the surface deformations of the El Asnam earthquake of October 10, 1980. *Tectonics,* 2: 17–49.

7. Hauksson, E., L.M. Jones, T.L. Davis, L.K. Hutton, A.G. Brady, P.A. Reasenberg, A.J. Michael, R.F. Yerkes, P. Williams, G. Reagor, C.W. Stover, A.L. Bent, A.K. Shakal, E. Etheredge, R.L. Porcella, C.G. Bufe, M.J.S. Johnston, and E. Cranswick, 1988. The 1987 Whittier Narrows earthquake in the Los Angeles metropolitan area, California. *Science,* 239: 1409–1412.

8. Bullard, T.F., and W.R. Lettis, 1993. Quaternary fold deformation associated with blind thrust faulting, Los Angeles Basin, California. *Journal of Geophysical Research,* 98: 8349–8369.

9. Davidson, K., 1994. Learning from Los Angeles. *Earth,* September: 40–47.

10. Zepeda, R.L., E.A. Keller, and T.K. Rockwell, 1991. Tectonic geomorphology of Wheeler Ridge. In E.A. Keller (ed.), Active Folding and Reverse Faulting in the Western Transverse Ranges, Southern California. *Geological Society of America Guidebook to 1991 Annual Meeting.* Geological Society of America: Boulder, CO.

11. Rich, J.L., 1934. Mechanics of low-angle overthrust faulting as illustrated by Cumberland thrust block, Virginia, Kentucky and Tennessee. *American Association of Petroleum Geologists Bulletin,* 18: 1584–1596.

12. Hamblin, W.K., 1965. Origin of "reverse drag" on the downthrown side of normal faults. *Geological Society of America Bulletin,* 74: 1145–1164.

13. Bruhn, R., 1994. Personal communication.

14. Sans de Galdeano, C., 1993. Personal communication (geologic maps); Granada, Spain.

15. Ikeda, Y., 1983. Thrust-front migration and its mechanism. *Bulletin of the Department of Geography, University of Tokyo,* (15): 125–159.

16. Yeats, R.S., 1986. Active faults related to folding. In *Active Tectonics.* National Academy Press: Washington, D.C.

17. Davis, D., J. Suppe, and F.A. Dahlen, 1983. Mechanics of fold-and-thrust belts and accretionary wedges. *Journal of Geophysical Research,* 88: 1153–1172.

18. Davis, T.L., 1983. Late Cenozoic structure and tectonic history of the western "Big Bend" of the San Andreas fault and adjacent San Emigdio Mountains. Ph.D. dissertation, University of California: Santa Barbara, CA.

19. Keller, E.A., R.L. Zepeda, D.B. Seaver, T.K. Rockwell, D.M. Laduzinsky, and D.L. Johnson, 1987. Active fold-thrust belts and the Western Transverse Ranges. *Geological Society of America Abstracts with Programs,* 19(6): 394.

20. Chapple, W.M., 1978. Mechanics of thin-skinned fold-and-thrust belts. *Geological Society of America Bulletin,* 89: 1189–1198.

21. Rockwell, T.K., E.A. Keller, M.N. Clark, and D.L. Johnson, 1984. Chronology and rates of faulting of Ventura River terraces, California. *Geological Society of America Bulletin,* 95: 1466–1474.

22. Rockwell, T.K., 1988. Neotectonics of the San Cayetano fault, Transverse Ranges, California. *Geological Society of America Bulletin,* 100: 500–513.

23. Keller, E.A., M.S. Bonkowski, R.J. Korsch, and R.J. Shlemmon, 1982. Tectonic geomorphology of the San Andreas fault zone in the southern Indio Hills, Coachella Valley, California. *Geological Society of America Bulletin,* 93: 46–56.

24. Lettis, W.R., and K.L. Hanson, 1991. Crustal strain partitioning: Implications for seismic-hazard assessment in western California. *Geology,* 19: 559–562.

25. Molnar, P., 1995. A review of active deformation of the Western Transverse Ranges (and its relevance to the active tectonics and earthquake history of Mongolia. *Abstracts from the SCEC Workshop: Thrust Ramps and Detachment Faults in the Western Transverse Ranges.* January 22–24, University of California, Santa Barbara: 16.

26. Dolan, J.F., K. Sieh, T.K. Rockwell, R.S. Yeats, J. Shaw, J. Suppe, G.J. Huftile, and E.M. Gath, 1995. Prospects for larger or more frequent earthquakes in the Los Angeles metropolitan region. *Science,* 267: 199–205.

27. Sylvester, A.G., R. Bilham, M. Jackson, and S. Barrientos, 1993. Aseismic growth of Durmid Hill, southeasternmost San Andreas fault, California. *Journal of Geophysical Research,* 98: 14,233–14,243.

28. Rockwell, T.K., E.A. Keller, and G.R. Dembroff, 1988. Quaternary rate of folding of the Ventura Avenue anticline, Western Transverse Ranges, Southern California. *Geological Society of America Bulletin,* 100: 850–858.

29. Medwedeff, D.A., 1992. Geometry and kinematics of an active, laterally propagating wedge thrust, Wheeler Ridge, California. In S. Mitra and G.W. Fisher (eds.), *Structural Geology of Fold-and-Thrust Belts.* Johns Hopkins University Press: Baltimore, MD.

30. Shelton, J.S., 1966. *Geology Illustrated.* W.H. Freeman: San Francisco.

31. Hall, N.T., 1984. Late Quaternary history of the eastern Pleito thrust fault, northern Transverse Ranges, California. Ph.D. dissertation, Stanford University, California.

32. Davis, T.L., and M.B. Lagoe, 1987. The Arvin-Tehachapi earthquake ($M = 7.7$) and its relationship to the White Wolf fault and the Pleito thrust system. *Geological Society of America Abstracts with Programs,* 19(6): 370.

8

Paleoseismology and Earthquake Prediction

PALEOSEISMOLOGY

Our ability to evaluate present and future earthquake hazards is based on understanding the past behavior of seismogenic (earthquake-producing) faults. Basic data for earthquake-hazard evaluation include the location, length, and amount of fault displacement; intensity of shaking; size or magnitude; and dates of previous earthquakes. This information is generally available for earthquakes in recent decades since local, regional, and global seismic networks of recording instruments have been used. To study historic earthquakes that predate the use of recording instruments, we rely on written accounts, coupled with geologic and geomorphic evidence. The historic record ranges from a few hundred years in the United States to a few thousand years in China. For prehistoric earthquakes, the only source of information is the geologic and geomorphic record [1].

Paleoseismicity is defined as the study of the occurrence, size, timing, and frequency of prehistoric earthquakes [1, 2]. Paleoseismicity, which most often utilizes the Pleistocene and Holocene (last 1.65 m.y. and last 10 k.y., respectively) geologic and geomorphic record, extends over a much longer time period than the limited historic record of seismic activity. This is particularly important because the recurrence intervals for large, damaging earthquakes on many fault segments in the most tectonically active regions on Earth are often a few hundred to a few thousand years. In regions with lesser tectonic activity, recurrence intervals may be tens- to hundreds-of-thousands of years [1].

EVIDENCE FOR PALEOEARTHQUAKES

Large earthquakes in the past have ruptured the surface of the Earth and folded near-surface Earth materials in response to subsurface rupture. Surface rupture and folding often can be discerned hundreds, thousands, or even tens-of-thousands of years later. Evidence for paleoseismic evaluation may be gathered from:

- Fault exposures
- Seismic-reflection profiles
- Faulted landforms such as stream terraces, offset streams, and marine terraces
- Fault scarps, including scarps produced by single earthquakes and composite scarps produced by several earthquakes
- Stratigraphic features such as colluvial wedges, liquefaction features, sand blows, fissure filling, and abrupt burial of deposits
- Folded alluvial deposits, rocks, and geomorphic surfaces

Fault Exposures

The best available scientific information from which to establish paleoseismicity is direct observation of faults. If we are able to study fault exposures directly, we may be able to measure displacements and/or collect material suitable for dating prehistoric earthquakes.

Some ways to examine faults include:

- **Natural exposures.** These include landforms such as seacliffs cut by wave erosion and a variety of steep slopes cut by stream and river processes. Natural exposures can be enhanced by clearing recent debris to better expose faulted deposits. A disadvantage of natural exposures is that they may not be in the most favorable orientation to a fault for paleoseismic evaluation.

- **Road cuts and railroad cuts.** Road cuts and railroad cuts are other good locations in which to search for fault exposures. Discovery of flexural-slip faults near Oak View, California (see Figures 2.25 and 7.19 and discussion in Chapter 7) began with one fortuitous observation in a highway cut.
- **Mines and gravel pits.** Gravel pits, in particular, provide important exposures because they typically are excavated in geologically young material (alluvium) and they may be relatively deep (tens of meters), exposing geologic features that are unlikely to be exposed naturally (see Figure 7.28).
- **Backhoe trenches.** Where neither naturally occurring nor preexisting exposures of a fault can be found, paleoseismologists may dig their own exposures using a backhoe or bulldozer (Figure 8.1). Sites for trenching are carefully chosen following preliminary geologic observation and

FIGURE 8.1
Excavation of a large fault trench, Point Conception, California.
(Photograph by E.A. Keller)

mapping. In order to evaluate both the horizontal and vertical components of displacement on strike-slip or oblique-slip faults (Figure 1.6), it is often useful to excavate two or more trenches. In addition to trenches across and perpendicular to the fault that allow the vertical component of displacement to be measured, one or more trenches are excavated parallel to the strike of the fault plane to evaluate the horizontal component of displacement. The objective is to identify a **piercing point,** which is a feature such as a buried channel, a buried pipe, or some other distinctive feature that is offset along the fault, so that it can be identified again where it exits the fault on the opposite side. Fault trenches excavated by bulldozers may be cut large enough to evaluate both horizontal and vertical components of displacement in a single trench. When datable material can be recovered from features with measured offset, then slip rates can be calculated as the ratio of displacement to the time during which the displacement has occurred. Of course, if an offset buried pipe is identified after a known earthquake, the only information obtained is the amount of displacement for that event. The best piercing points from which to calculate a slip rate are features several thousand to several tens of thousands of years old; that is, features that are old enough to have cumulative offset from numerous earthquakes, providing a longer term average slip rate.

- **Boreholes.** In addition to trenches, boreholes may be drilled by a variety of techniques to gain information about a fault. Boreholes vary from small-diameter holes from which a continuous core (cylindrical section of earth material a few centimeters in diameter) is obtained to large-diameter boreholes that a person can go down into, protected by a special cage, for direct observation. Boreholes are effective ways to identify and locate faults, but because of limited exposure, they may not provide sufficient information for detailed paleoseismology investigations. They can, however, provide important information concerning the vertical component of fault displacement and help identify sites suitable for trenching.

Seismic Reflection

Seismic-reflection profiling is another method for gaining information about fault offsets without physically excavating or drilling. In this technique, the ground is shaken by a small explosive or a mechanical vibrator at one source (a **shotpoint**), the vibration travels beneath the surface as waves and is detected at a series of receivers (**geophones**) (Figure 8.2). The path and travel time of the vibrations are sensitive to density contrasts of the underlying rocks. Through a computerized process of decoding the seismic vibration, an image of the rocks beneath the surface is created, including any disruptions of the sequence caused by fault offsets. Figure 8.3 illustrates a shallow seismic-reflection profile across

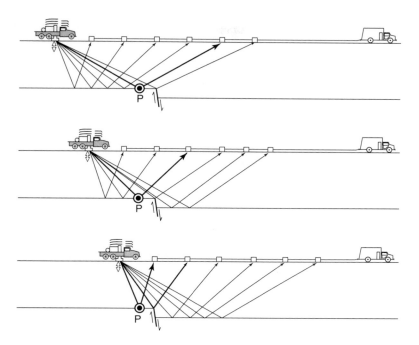

FIGURE 8.2
Idealized diagram showing how shallow seismic-reflection profiling can be used to iden-
tify a fault.
(From Cook et al., 1980. *The Southern Appalachians and the Growth of Continents.* Scientific American,
Inc. All rights reserved.)

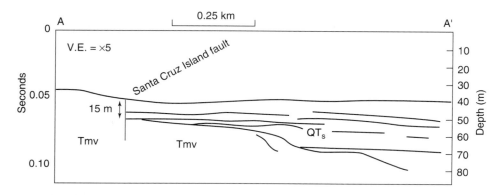

FIGURE 8.3
Interpretative cross section made from shallow seismic data across the Santa Cruz Island
fault. This diagram suggests that there are approximately 15 m of vertical displacement of
the Tertiary volcanic rock section *(Tmv)*. *QTs* is Quaternary marine terrace deposits.
(From Pinter and Sorlien, 1991. *Geology,* 19: 909–912)

the Santa Cruz Island fault, California. The profile was created by a ship towing a vibration source and an array of geophones. Following data processing, the data suggest a 15 m vertical separation between volcanic rocks and recent marine sediments, probably caused by the last two or three ruptures on that fault. A related method is **ground-penetrating radar.** Electromagnetic reflections from a radar source are decoded in a manner similar to that for seismic reflection.

Faulted Landforms

A landform created prior to the last surface-rupturing earthquake at a site may preserve a record of the rupture, such as:

- **Stream terraces.** Terraces consist of broad surfaces, sometimes datable, that can be particularly useful in studying paleoearthquakes (see Chapters 2 and 5, and Figures 2.25, 2.26, and 5.19). Figure 8.4 is an idealized block diagram that shows two possible scenarios of how active faulting might displace river terraces. These two scenarios illustrate that trying to determine the pattern of displacement of river terraces may be a difficult problem. Sometimes it may be difficult to tell if the scarps present are fault scarps or erosional scarps produced from river processes.
- **Offset streams.** Strike-slip faults with little or no vertical component of motion will not cause large vertical deformation of terraces or subhori-

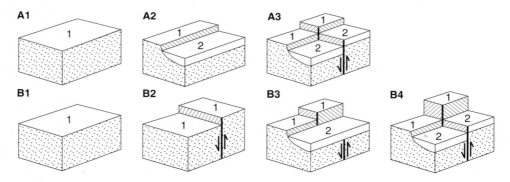

FIGURE 8.4
Idealized block diagrams illustrating potential complexities of interpreting river terraces that have been faulted. Sequence (A) shows the development of two river terraces (1, 2) that are subsequently faulted (A3). Sequence (B) is more complex. Terrace 1 is faulted (B2). Following faulting, terrace 2 forms (B3), and finally the sequence is faulted again (B4). Because faulting occurred at two specific times, the fault scarp for terrace 1 is higher than that for terrace 2. This illustration (B4) shows a multiple-event scarp on terrace 1. (From McCalpin, 1987 [11])

zontal surfaces of other landforms. However, streams that cross a strike-slip fault may be laterally offset (see Chapters 2 and 5, and Figure 2.4). Figure 8.5 shows several offset or deflected streams along the Santa Cruz Island fault in Southern California. This fault is left-lateral and so the streams are displaced to the left—if you were to follow the stream, either upstream or downstream, you would have to turn to the left to follow the channel across the fault.

- **Marine terraces.** Like stream terraces, the broad, subhorizontal surfaces of marine terraces are useful for measuring vertical fault motions and estimating the age of paleoearthquakes (see Chapter 6). Where a series of marine terraces of different ages are all displaced by a single fault and the older terraces are displaced more than younger ones, we may infer several ruptures on the fault [3] (Figure 8.6).

Uplift during an earthquake in a coastal area may produce a single marine terrace. Dating the terrace then establishes when that earthquake occurred. Sometimes, along a tectonically-active coastline, a series of uplifted terraces, each produced by a separate earthquake, may be present. This phenomenon has been observed in California, Alaska, New Zealand, Japan, and elsewhere. Figure 8.7A

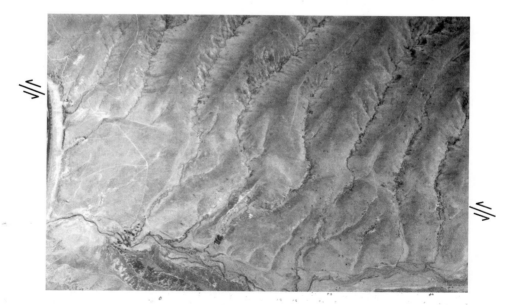

FIGURE 8.5
Aerial photograph of the Santa Cruz Island fault zone (roughly E to W, right central to left central between arrows) showing several offset and deflected streams (central and right central). The streams are offset by left-lateral strike-slip motion on the fault. Distance across the photograph is about 3 km.
(Photograph courtesy of Pacific Western Aerial Survey, Santa Barbara, CA)

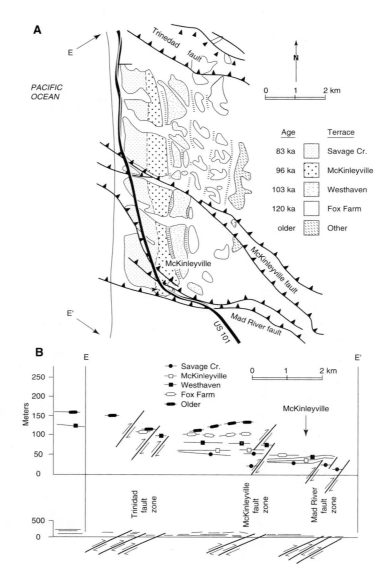

FIGURE 8.6

(A) Reverse faults associated with the Cascadia subduction zone displace a series of terraces. Note that where the terraces cross the McKinleyville fault zone (B), the 120 ka Fox Farm terrace is displaced approximately 50 m vertically, whereas the Westhaven terrace (estimated age of 103 ka) is offset approximately 30 m; the 96 ka McKinleyville terrace is offset approximately 20 m; and the 83 ka Savage Creek terrace is offset <20 m. These data suggest that the rate of uplift along the fault system is approximately 0.2–0.4 mm/yr. However, because these terraces cross several fault zones, including the Little River fault zone to the north and the Mad River fault zone to the south, the total rate of uplift is closer to 1 mm/yr [3]. This example from the Cascadia Subduction Zone illustrates how faulted marine terraces provide information concerning rates of uplift, as well as how uplift is partitioned among several fault zones.

(From Carver and Burke, 1992 [3])

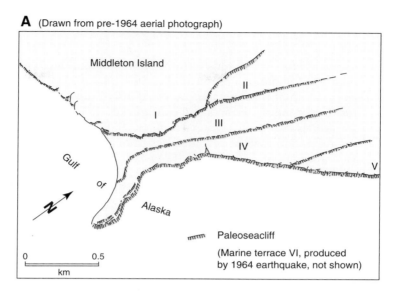

A (Drawn from pre-1964 aerial photograph)

Middleton Island

II

I

III

IV

V

Gulf

of

Alaska

🔺 Paleoseacliff

(Marine terrace VI, produced
by 1964 earthquake, not shown)

0   0.5

km

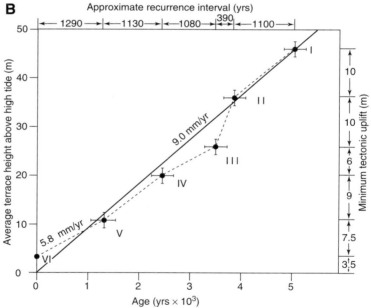

B   Approximate recurrence interval (yrs)

|← 1290 →|← 1130 →|← 1080 →|390|← 1100 →|

FIGURE 8.7
(A) Drawing from an aerial photograph (1947) of the southeastern end of Middleton
Island, Alaska. The island emerged approximately 4.9 ka, and the terraces were each
formed by 5 m to 7 m of coseismic uplift. The 1964 earthquake caused additional uplift of
approximately 3.4 m, forming a sixth terrace. (B) Graph showing the uplift of all six terraces. Average rate of uplift is 9.0 mm/yr with uplift per event varying from approximately 3.5 m to 10 m. Approximate recurrence intervals between earthquakes varied
from about 400 yr to 1300 yr.
(From Plafker, 1987 [4])

shows a series of five such terraces on Middleton Island, Alaska [4]. All five terraces on a 1947 photograph predate the 1964 ($M = 8.25$) earthquake which uplifted the coast 3.5 m at that site and produced a sixth terrace. Figure 8.7B shows the uplift history for the six terraces and earthquakes that produced them. Chronology of the terraces is based on carbon-14 dates. Note that the average uplift rate is about 9 mm/yr, but uplift per event varies from about 4 m to 10 m, and recurrence intervals from about 400 yr to 1300 yr [4]. All five prehistoric events (I to V, Figure 8.7A) are thought to have been great earthquakes ($M > 8$) because they produced more uplift than did the 1964 event. Thus, to summarize the paleoseismic activity at Middleton Island, there have been six great earthquakes in this region in the last 5 k.y., with average recurrence intervals (with one exception) of about 1 k.y.

Fault Scarps

Fault scarps are the direct manifestation of surface-rupturing earthquakes. They are produced almost instantaneously as an earthquake rupture propagates to the surface. People who have observed earthquakes firsthand commonly state that the scarps and other fractures form very quickly, racing across the landscape like a giant zipper being undone [5]. Figure 8.8 shows a fault scarp over 2 m high that formed during the 1992 ($M = 7.5$) earthquake near Landers, California. In normal or reverse faulting, the orientation of the scarps indicates the direction of slip, but in strike-slip faulting, scarps face in different directions (see Chapter 2).

Fault scarps are slopes and, as such, have a basic morphology common to many natural slopes (Figure 8.9). Not all of the slope elements shown on Figure 8.9 may be present on a given fault scarp; in fact, the dominance of one element over another will change with time. It is important to recognize that the different elements of a fault scarp are produced by different processes. For example, the free face, produced directly by faulting, may be nearly vertical when it forms. On the other hand, the debris slope and wash slope are related to accumulation of material at the base of the free face, and thus are associated with erosional and sediment-transport processes. **Fault-scarp degradation** proceeds at variable rates, depending upon climatic conditions and the types of materials in which the scarp formed [6]. Changes in slope elements through time can be recorded as a percent of the scarp length; for example, a scarp might be composed of a free face over roughly 50% of the profile, and a debris slope over the remaining 50%. Examination of Figure 8.9 suggests some interesting aspects of fault-scarp degradation:

- The free face is a steep to vertical slope formed directly by rupture of the surface by an active fault. In loose, unconsolidated material, the free face would instantaneously collapse, but soil moisture and cohesion of the

FIGURE 8.8
Fault scarp produced by the 1992 Landers earthquake ($M = 7.5$). The scarp is approximately 2 m high.
(Photograph by E.A. Keller)

FIGURE 8.9
Basic slope elements that may be present on a fault scarp.
(From Wallace, 1977 [6])

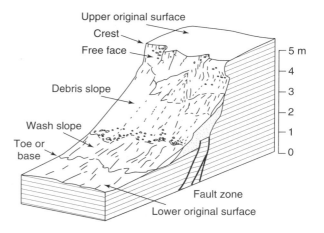

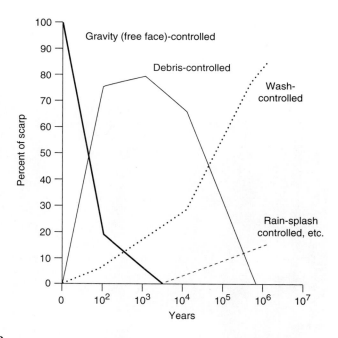

FIGURE 8.10
Diagram showing change in slope elements (fault-scarp morphology) through time for fault-scarp degradation in the Basin and Range.
(From Wallace, 1977 [6])

    material hold the slope together for a period that can range from a few decades to several thousand years.
- The debris slope results from material deposited at the base of the free face by gravity. It starts out as a very small portion of the slope, but increases to dominate the scarp within a few thousand years. Later, the dominance of the debris slope wanes in favor of the wash slope.
- The wash slope is produced by deposition of material near the toe of the slope, usually by running water. This slope element is less steep than the debris slope, although there often is no abrupt boundary between the two elements. The wash slope begins to form slowly, but eventually dominates the morphology of the scarp after a period of tens to hundreds of thousands of years. Therefore very old scarps will have a gentle slope and consist primarily of wash slope.

    Fault scarps have been studied intensively in the Basin and Range province of Nevada (Figure 8.10). Table 8.1 summarizes fault-scarp morphol-

TABLE 8.1
Fault scarp-slope morphology.

Slope Element	Morphology	Process (Formation and/or Modification)	Comments and General Chronology
Crest	Top of fault scarp (break in slope) initially sharp, becomes rounded with time	Produced by faulting; modified by weathering, mass wasting	Becomes rounded after free face disappears; usually rounded after about 10,000 yr
Free Face	Straight segment; initially 45° to overhanging	Produced by faulting; modified by weathering, gullying, mass wasting; eventually buried from below by accumulation of debris	Dominant element for 100 years or so; disappears after about 1000–2000 yr
Debris Slope	Straight segment; angle of repose of material usually 30° to 38°	Accumulation of material that has fallen down from the free face	Is dominant element after about 100 yr, remains dominant until about 100,000 yr, disappears at about 1,000,000 yr
Wash Slope	Straight to gently concave segment; overlaps the debris slope; slope angle generally 3° to 15°	Erosion and deposition by water; deposition of wedge or fan of alluvium near toe of the slope; some gullying	Is developed by 100 yr, significant by 1000 yr, and dominant by 100,000 yr
Toe	Base of fault scarp (break in slope); may be initially sharp, but with time may become indeterminate as grades into original slope	Erosion and deposition by water; owing to change in process/form from up-slope element (free face, debris slope, or wash slope) to original surface below the fault scarp slope	More prominent in young fault scarps or where wash slope is not present; on scarps older than about 12,000 yr, the basal slope break is sharper than the crestal slope break

(After Wallace, 1977 [6])

ogy and change with time for fault scarps in the Basin and Range [6]. The chronology of fault-scarp degradation in the Basin and Range was developed by studying fault scarps that truncate features of known ages such as Pleistocene lake shorelines dated by the carbon-14 method. In other cases, fault scarps are associated with volcanic ash deposits or are dated by tree rings (dendrochronology) [6].

FAULT-SCARP DEGRADATION

The processes that govern the morphology and evolution of slopes formed on alluvial materials include erosion, deposition, and sediment transport. It has been recognized that these processes can be modeled quantitatively. Fault scarps are a type of slope that is particularly well suited to numerical models, because scarps form instantaneously during an earthquake and then begin to degrade. The most common application of scarp-degradation models is to estimate the age of fault scarps. In particular, models based on the **diffusion equation** are used to answer this question [7, 8].

The diffusion equation is expressed as follows:

$$\frac{\partial z}{\partial t} = k \frac{\partial^2 z}{\partial x^2} \tag{8.1}$$

where z is elevation of a point on the slope, x is horizontal distance, t is time, and k is a constant known as **diffusivity** (Figure 8.11), which depends on climate and material. Equation 8.1 simply states that the rate of erosion or aggradation at a given point is proportional to the curvature of the profile at that point; that is, the sharpest corners on the profile will tend to change, to be smoothed out fastest.

The diffusion equation comes from physics, where it is used to describe the flow of heat across a thermal boundary. The analogy of heat flow makes diffusion degradation of fault scarps a bit clearer (Figure 8.12). Elevation on the scarp profile, z, is equivalent to temperature in the physics problem. The conditions at $t = 0$ are of an abrupt juxtaposition of high temperature against low temperature at the thermal boundary. Just as heat flows from the region of high temperature to low, sediment is transported from the region of high elevation to low elevation on a degrading slope. The overall rate of flow is determined by the

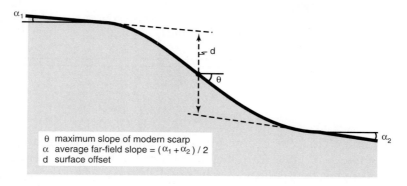

θ maximum slope of modern scarp
α average far-field slope = $(\alpha_1 + \alpha_2) / 2$
d surface offset

FIGURE 8.11
Idealized fault scarp, illustrating parameters that can be directly measured in the field. Numerical values of these parameters are utilized in a solution of the diffusion equation to estimate the age of the scarp, and thus when the earthquake occurred.

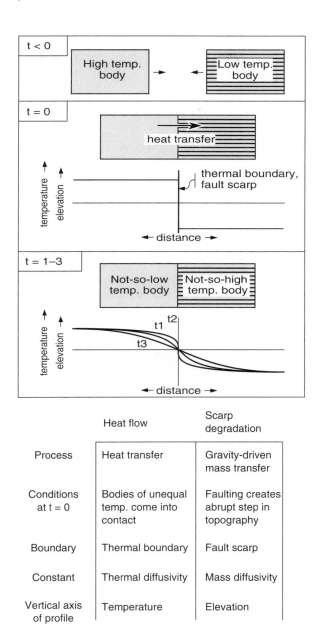

	Heat flow	Scarp degradation
Process	Heat transfer	Gravity-driven mass transfer
Conditions at t = 0	Bodies of unequal temp. come into contact	Faulting creates abrupt step in topography
Boundary	Thermal boundary	Fault scarp
Constant	Thermal diffusivity	Mass diffusivity
Vertical axis of profile	Temperature	Elevation

FIGURE 8.12

Heat transfer from a warm mass to a cold one illustrates the principle of diffusion, which also governs fault-scarp degradation. Just as heat flows from the hot material (high thermal potential) to the cold material (low thermal potential) in thermal diffusion, sediment erodes from the top of a slope (high gravitational potential) and is deposited at the base (low gravitational potential).

conductive properties of the media in the case of heat flow, and by properties of erosivity (climate) and erodability (texture) in the case of sediment transport. The result in both cases is that the profile becomes increasingly subdued with time.

Applying a diffusion model or a solution of the diffusion equation to a fault scarp requires several prerequisites and assumptions:

- the slope must be transport-limited
- the scarp must have formed instantaneously in a single earthquake
- the sediment must be assumed to be nearly cohesionless

Transport-limited refers to a situation where there is an abundant supply of loose sediment, and degradation of the slope is

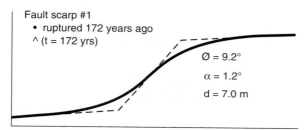

Fault scarp #1
- ruptured 172 years ago
- ^ (t = 172 yrs)

$\emptyset = 9.2°$

$\alpha = 1.2°$

$d = 7.0$ m

Using equation 8.2:

$$k \ (172 \ yr) = \frac{(7.0 \ m)^2}{4\pi} \ \frac{1}{[(\tan(9.2°) - \tan(1.2°)]}$$

$$k = 0.16 \ m^2/yr$$

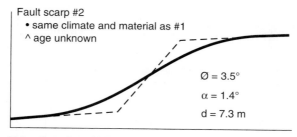

Fault scarp #2
- same climate and material as #1
- ^ age unknown

$\emptyset = 3.5°$

$\alpha = 1.4°$

$d = 7.3$ m

Using the value of k from scarp #1:

$$(0.16 \ m^2/yr) \ t = \frac{(7.3 \ m)^2}{4\pi} \ \frac{1}{[(\tan(3.5°) - \tan(1.4°)]}$$

$$= 722 \ yr$$

FIGURE 8.13

Idealized diagram showing two fault scarps. For fault scarp 1, the date of the earthquake is known and diffusivity, k, is calculated. This value of diffusivity (k) is used for fault scarp 2 to estimate that the earthquake occurred approximately 722 yr ago.

controlled only by the dynamics of the processes of sediment transport. Slopes that are controlled by the rate of weathering ("weathering-limited"; for example, bedrock scarps) are fundamentally different. The second requirement is that the scarp formed in one earthquake, and was not built by several small offsets on the same fault. Third, diffusion models generally assume that the free face of the scarp crumbled almost instantaneously after the earthquake event and reached the angle of repose.

Given these assumptions, fairly simple solutions to the diffusion equation are possible. For example [7]:

$$k\,t = \frac{d^2}{4\pi}\;\frac{1}{(\tan\theta - \tan\alpha)} \qquad (8.2)$$

where d is vertical displacement, θ is maximum scarp angle, and α is far-field slope (Figure 8.11). Note that solutions like Equation 8.2 always include both diffusivity and time;

neither of those parameters can be solved independently of the other. On a scarp of known age, a profile measured across it allows us to determine the value of diffusivity. Where two fault scarps occur in approximately the same climate and material, one scarp of known age and the other of unknown age, we can calculate the value of diffusivity on the dated scarp and use it to estimate the age of the undated scarp (Figure 8.13).

Since mass diffusion was introduced as a model for fault-scarp degradation, several complications and refinements have been proposed. For example, in addition to lithology and regional climate, the value of diffusivity can be affected by slope orientation, creating variability between north- and south-facing scarps even in the same vicinity [9]. It has also been found that the gradient of the surface away from the effects of the fault, or "far-field slope," can have first-order effects on age determinations by the diffusion equation [10].

Stratigraphic Evidence for Earthquakes

Stratigraphic evidence for paleoseismicity is one type of **event stratigraphy.** This name is used because we are discussing features found in stratigraphic sequences that indicate past and rare events; in this case our concern is past earthquake events. These features include displaced strata, colluvial wedges, sand boils, fissure filling, and abrupt burial.

Displaced Strata. The clearest evidence of past earthquakes found in fault exposures is displaced strata. Often when a trench is exposed or a natural exposure is examined, there may be several fault strands present. Some of these will displace older material and be buried by younger material, whereas other nearby fault strands cut the entire sequence. Careful evaluation of the displacement history can help establish the number of faulting events that have occurred. For example, Figure 8.14 shows an idealized diagram of sand and gravel deposits that have been faulted by three splays or strands of a fault system. Fault 1 cuts unit C and has left a buried fault scarp between B and C; fault 2 cuts units B and C and has left a buried fault scarp between A and B; and

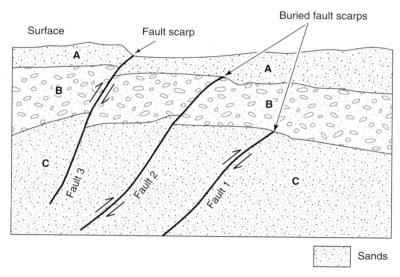

FIGURE 8.14
Trench exposure showing displacement of sand and gravel deposits, buried fault scarps, and a surface fault scarp. Fault 1 displaces only unit C. Fault 2 displaces B and C, and fault 3 displaces A, B, and C. This stratigraphy, along with buried fault scarps and the surface fault scarp, suggests that three discrete faulting events occurred. The oldest faulting event occurred on fault 1 and the youngest on fault 3.

fault 3 cuts units A, B, and C and has a fault scarp at the surface. One way to interpret this stratigraphy is that there have been three earthquake events: the oldest is fault 1, followed by rupture on fault 2, and the youngest rupture is on fault 3. Often the stratigraphy exposed in a fault trench is much more complicated, with sedimentary units folded and faulted in a complex way. Complex stratigraphy requires very careful work to determine the history of faulting events.

Colluvial Wedges. **Colluvium** is unconsolidated material found at the base of steep slopes. Colluvial deposits generally are more angular than stream deposits, not having been transported very far. Following an earthquake which produces a fault scarp, a colluvial wedge may form at the base of the fault scarp as the free face degrades to a debris slope. Sometimes these colluvial wedges are buried and preserved in the stratigraphic record as evidence of past earthquakes (Figure 8.15). As Figure 8.15 illustrates, the colluvial wedges may be rotated toward the fault as displacement continues. There may also be some soil development on top of the colluvial wedge, because there often is sufficient time between events for a weak soil to form [11]. If fault-scarp colluvial wedges can be recognized in exposures and dated, this information is valuable in working out the timing of past earthquake events.

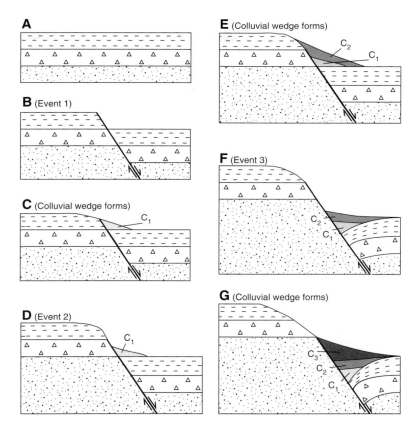

FIGURE 8.15
Development of a three-event fault scarp. Each faulting event is followed by the genera-
tion of a fault-scarp colluvial wedge (C_1, C_2, and C_3).
(After McCalpin, 1987 [11])

Reverse faults also produce colluvial wedges. As a reverse fault ruptures
the surface, it has an oversteepened front that quickly collapses, forming a colluvial wedge (Figure 8.16). The hanging wall may be warped, forming **drag folds,**
so called because they are dragged (folded) by displacement along the fault
plane. For locating fault trenches, experience has shown that reverse faults are
likely to be found about halfway up the scarp. Material quickly collapses from
the leading edge of a reverse fault, depositing colluvium below the fault as
quickly as the area above is uplifted.

Sand Boils. Sand-boil deposits, sometimes also called "sand craters," have been
associated with many earthquakes. At the surface, they are characterized by low
mounds of sand that have been extruded from fractures. Layered sedimentary
deposits beneath the surface may, during an earthquake, experience high fluid

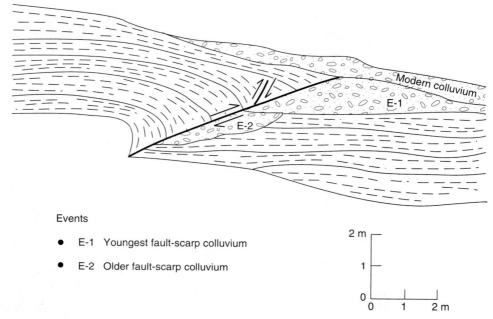

Events

● E-1 Youngest fault-scarp colluvium

● E-2 Older fault-scarp colluvium

FIGURE 8.16
Idealized and simplified fault-trench log for the McKinleyville fault, Humboldt County,
California, showing two events of faulting and formation of fault-scarp colluvium.
(From Carver, 1987. In A.J. Crone and E.M. Omdahl (eds.), Directions in Paleoseismology. *U.S. Geological Survey Open-File Report* 87-673)

pressure and liquefaction. Fluidized sand is squeezed to the surface, forming a
thin, circular deposit of sand (Figure 8.17A). Evidence of sand boils in a
stratigraphic sequence (Figure 8.17B) represents liquefied sand extruded onto
the surface and later buried by other materials. When such deposits are
identified in fault exposures, they are evidence of a past earthquake event, but
they are not conclusive evidence because sand boils also can form without
earthquakes (for example, by discharge of sediment-laden runoff routed from
preexisting fissures) [12].

Fissure Filling. Large earthquakes may form numerous fissures and cracks
(Figure 8.18). These are unlikely to remain open for very long because the sides
of the fissures and cracks are very steep. Material from the surface and from the
sides of the fissures soon fills them. Eventually new material may be deposited
over the filled fracture. When these features are recognized in fault exposures,
they also are evidence of past earthquakes. As with sand boils, fissures can form
without earthquakes (for example, due to landsliding or groundwater
withdrawal).

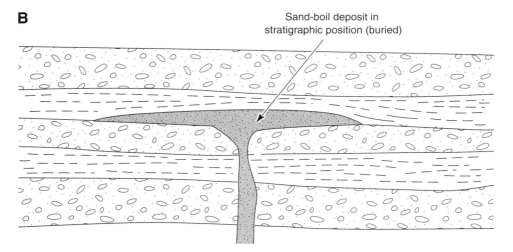

FIGURE 8.17
(A) Sand boil produced by the 1989 Loma Prieta ($M = 7.1$) earthquake. Note street curb for scale. (B) Idealized diagram showing how a sand boil may appear in the stratigraphic record following burial. It is important to keep in mind, however, that sand boils are also produced by processes other than earthquakes.
(Photograph courtesy of D. Laduzinsky)

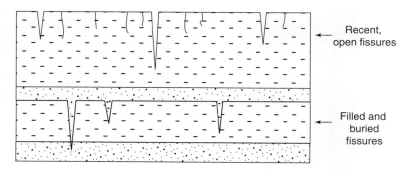

FIGURE 8.18
Recent, open fissures and older, filled and buried fissures. As with sand boils, fissures are not absolute proof of earthquakes because they may be produced by several other processes as well as earthquakes.

Abrupt Burial. Figure 8.19 shows a "ghost forest," the trees of which were killed suddenly by a subsidence event approximately 300 yr ago. It has been suggested recently that abrupt burial of forest or of salt-marsh deposits may be characteristic of large subduction-zone earthquakes along the Cascadia Subduction Zone (also see discussion in Chapter 6) [13]. The hypothesis is that during great subduction-zone earthquakes, parts of the coast may subside by a meter or so. Until recently it was thought that no great earthquakes occurred along the Washington and Oregon coasts; however, the discovery of the abrupt burial of salt-marsh and forest environments suggests that this may not be the case. In several locations from Northern California to Washington there is evidence of two or more episodes of abrupt burial in the past few thousand years. These locations are being studied intensely to determine if, in fact, the burial was rapid and coseismic.

FAULT-ZONE SEGMENTATION

The concept of fault-zone segmentation has been recognized for over 20 years [14]. The basic idea is that for long faults, an earthquake seldom ruptures along the entire length of the fault. More commonly, only one, or perhaps two, segments rupture during a large event. For example, on a regional scale, the San Andreas fault of California is subdivided into four segments (Figure 8.20) [6]. Segment boundaries are based on the historical behavior of the fault. The south-central segment ruptured in 1857, producing a great earthquake. The northern segment ruptured in 1906. The southern segment has not produced a major earthquake ($M > 7$) in historic time.

The simple regional segmentation of the San Andreas fault (Figure 8.20) might lead you to believe that the concept of fault segmentation is simple.

FIGURE 8.19
Photograph of the "Ghost Forest" on the coast of western Washington. It is hypothesized that these trees were killed approximately 300 yr ago by a giant earthquake that caused subsidence, submerging the trees below sea level.
(Photograph courtesy of United States Geological Survey and B. Atwater)

Unfortunately, this is not the case. There is general agreement that faults may be segmented at a variety of scales (from a few meters to several tens of kilometers in length) [15]. In addition, segmentation based on historic activity is only one approach. Structural segmentation can be recognized by changes in fault-zone geomorphology or fault-trace orientation such as bends, step-overs and separations or gaps [16]. **Structural segmentation** also occurs where segments intersect with other faults or folds. The end of a segment is a structural discontinuity. Finally, structural segmentation can also result from changes in geologic materials along a fault zone or local heterogeneities along a fault plane [15, 16].

The most basic approach to fault-zone segmentation is to define **earthquake segments** based on rupture behavior during earthquakes. Earthquake segmentation may be determined from historic earthquakes or by paleoseismic evaluation. An earthquake segment is defined as those parts of a fault zone that rupture as a unit during an earthquake [17]. Earthquake segmentation is the

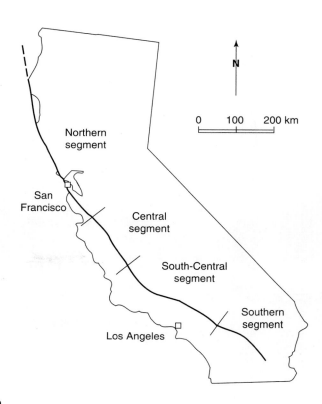

FIGURE 8.20
Simplified segmentation model for the San Andreas fault zone, California. The northern
segment ruptured in the 1906 San Francisco earthquake; the central segment is characterized by relatively frequent modern earthquakes and creep; the south-central segment
most recently ruptured in a great earthquake in 1857; and the southern segment has not
ruptured in historic times.
(From Schwartz and Coppersmith, 1986 [19])

most important concept for evaluating seismic hazard, but a better understanding of all types of fault segmentation is necessary to better understand earthquake mechanics, including why and where ruptures start and terminate [15].
Of particular importance for large earthquakes ($M > 7$) is the need to know how
rupture of multiple fault segments occurs during earthquakes [16].

The concept of fault segmentation is important because it has implications for:

- Long-term earthquake forecasting involving probabilistic assessment of
  seismic hazard (see Chapter 1). Conditional-probability analysis requires
  information on slip rate, slip per event, and recurrence interval of earthquakes for specific fault segments.
- Estimating the maximum earthquake likely to occur on a particular fault
  [15] (see Figure 1.27).

- Estimating ground motion produced by an earthquake. It is believed that rupture propagation is related to fault-zone structure, and thus segmentation [15].
- Identifying areas along a fault zone where earthquakes nucleate (rupture starts), as well as areas that may act as barriers to earthquake rupture (where rupture ends) [16].
- Better understanding the fundamental mechanisms associated with earthquake generation and faulting.
- Better understanding the complexities of large, damaging earthquakes that produce ruptures along several geometric and structural segments of a fault zone. This is particularly important because it is a common practice to map a particular fault zone by linking two or more segments together. As total length of a fault zone increases, the magnitude of the largest probable earthquake increases.

CASE STUDY

SEGMENTATION AND PALEOSEISMICITY OF THE WASATCH FAULT ZONE, UTAH

The Wasatch Fault Zone (WFZ) is one of the most-studied normal fault zones in the western United States. Extending nearly 350 km from southern Idaho southward past Great Salt Lake and Salt Lake City, it is also one of the longest. Paleoseismic evaluation of the WFZ resulted in the development of the concept of characteristic earthquakes (see Chapter 1) as well as an improved understanding of fault-zone segmentation [17, 18, 19].

Although there have not been historic earthquakes on the Wasatch Fault Zone (Figure 8.21), there is paleoseismic evidence that it has ruptured several times in the Holocene (last 10 k.y.); the most recent event was about 400 yr ago [17, 18, 19]. The WFZ is generally coincident with the boundary between the Wasatch Mountains and the adjacent basin in which Great Salt Lake and Utah Lake are located (see Figure 8.21). Proposed fault segments for the WFZ

(Figure 8.21) have boundaries that generally coincide with geometric or structural changes such as changes in surface trend of the fault zone; major salients (a salient is defined structurally as a block of bedrock that extends into the basin, or geomorphically as a landform that extends outward from surrounding topography); and other geologic and structural discontinuities, such as older thrust faults or cross faults [19]. There has been discussion of how many segments are present in the WFZ [17, 18], but there is general agreement that segmentation is very important in characterizing the past earthquake activity and evaluating the seismic hazard.

During the past 25 yr, over 40 trenches have been excavated to gather paleoseismic data at 18 sites along the segments of the WFZ. Table 8.2 summarizes some of the information obtained from trench studies and studies of natural fault exposures.

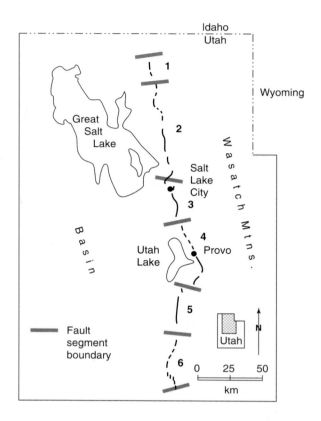

FIGURE 8.21
Map of the Wasatch Fault Zone, Utah, showing six fault segments. The fault system has normal displacement, with the upthrown side being the Wasatch Mountains. Dark bands define boundaries to fault segments and generally correspond to geometric and structural discontinuities along the fault zone.
(After Schwartz and Coppersmith, 1986 [19])

These data [17, 18, 19] for the WFZ suggest:

- The northern (Collinston) segment has not been active in the Holocene.
- Assuming that each prehistoric event ruptured most of a particular segment (30 km to 70 km), past earthquakes range from large to major ($M = 6.5$ to $M = 7.7$). These magnitudes are consistent with historic earthquakes in the Basin and Range.

- Segments 2–5 have experienced multiple large to major ($M = 6.5$ to $M = 7.7$) earthquakes during the Holocene, some in the past few hundred years.
- There is no evidence for a major ($M = 7.7$) earthquake on the WFZ in the past 400 yr.
- During the past 5.5 k.y., there has been a large to major earthquake ($M = 6.5$ to $M = 7.7$) about every 400 yr. somewhere on segments 2–6.

TABLE 8.2
Paleoseismic evaluation for the six segments of the Wasatch fault zone shown in Figure 8.21.

Segment	Length (km)	Approximate displacement per event (m)	Approximate slip rate (mm/yr)	Approximate recurrence interval (yr)	Comment
1	30	–	–	>10,000	No known surface displacement past 13,500 yr
2	70	1.6	1.3 (+0.5, –0.2)	1000 to 1500	4 Holocene events. Most recent about 500 yr ago, oldest 4000 yr
3	35	2.0	0.76 (+0.6, –0.2)	1500 to 3500	2 Holocene events. Most recent about 1500 yr ago, oldest 5000 yr
4	55	1.6 to 2.3	0.85 to 1.0	1500 to 3000	3 Holocene events. Most recent about 500 yr ago, oldest 5000 yr ago
5	35	2.3	1.27 to 1.36	1500 to 2000	3 Holocene events. Most recent about 400 yr ago
6	40	2.5	less than 0.35	7000	1 Holocene event about 1000 yr ago

(Data from Schwartz and Coppersmith, 1984 [18]; Schwartz and Coppersmith, 1986 [19]; and Machette et al., 1989 [17])

- The central portion of the fault zone (segments 2–5) is more active (more events, higher slip rates, and lower recurrence interval) than the northern or southern ends (segments 1 and 6).
- Approximate Holocene recurrence intervals on each of the more active center segments (segments 2–5) vary from about 1 k.y. to 4 k.y.

Evaluation using detailed age control of faulting on the WFZ [17] indicates information concerning temporal clustering of earthquakes:

- During the past 1 k.y. there has been one major earthquake about every 200 yr somewhere on segments 2–6.
- There is no evidence for strong clustering of earthquakes between about 1 ka and 5.5 ka.

In summary, paleoseismic evaluation of WFZ suggests that large, damaging earth-quakes have occurred on several fault segments in the past few thousand years. The most recent events occurred 400 to 500 yr ago. In the past 1 k.y., there has been a temporal clustering of events on the central segments (2–5), with large to major earthquakes ($M = 6.5$ to $M = 7.7$) about every 200 yr. This paleoseismic information is critical for developing strategies to minimize future earthquake damage to urban areas such as Salt Lake City and Provo, Utah.

Study of the Wasatch Fault Zone supports the **characteristic earthquake** concept (see Chapter 1). This concept implies that a given fault segment has particular geomechanical properties that lead to repeated earthquakes on that segment through time with approximately equal magnitudes and amounts of displacement [18]. Some investigators have used the characteristic-earthquake model to infer that the recurrence time or recurrence interval is approximately the same as well; this would follow if (1) the

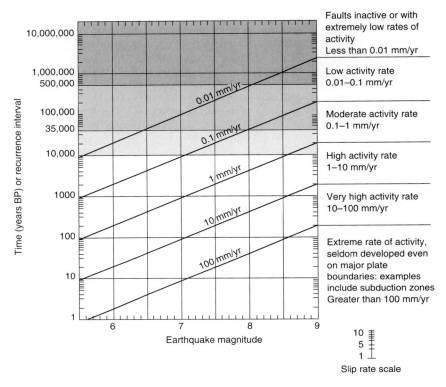

FIGURE 8.22
Relationships between recurrence interval, slip rate, and earthquake magnitude.
(After Slemmons and Depolo, 1986. In *Active Tectonics*. National Academy Press: Washington, D.C.)

slip rate of the fault, (2) the magnitude of the earthquake, and (3) the displacement per event were all constant. Under these circumstances, the average recurrence time is equal to the ratio of displacement per event to the slip rate [20]. Figure 8.22 relates the three parameters above, based on the assumption that a $M = 7$ earthquake will produce a 1 m characteristic displacement, a $M = 8$ will produce 5 m, and a $M = 9$ will produce 20 m. Figure 8.22 does appear to provide "ballpark" figures for recurrence intervals of earthquakes. However, when we examine the history of a particular fault and have sufficient data to be able to estimate

slip rates and recurrence intervals, it is not uncommon to find that slip rate and time between events vary considerably for different fault segments, as illustrated by the WFZ example. The more we learn about earthquakes, the more it becomes apparent that there is a general lack of uniformity in earthquake frequency for a particular fault zone composed of several segments. Nevertheless, although average slip rates and recurrence intervals have limitations in predicting the magnitude and time of the next rupture, they do provide useful guidelines for land-use planning, building codes, and engineering design [21].

CASE STUDY

TWELVE CENTURIES OF EARTHQUAKES ON THE SAN ANDREAS FAULT, SOUTHERN CALIFORNIA

Undoubtedly one of the most remarkable sites for paleoseismic studies in the world is Pallett Creek, located approximately 55 km northeast of Los Angeles. The site contains evidence of ten large earthquakes, two of which occurred in historical time (1812 and 1857). High-precision radiocarbon analyses provide accurate dating of most of the eight prehistoric events, extending back to approximately A.D. 671. (Figure 8.23) [22]. When the

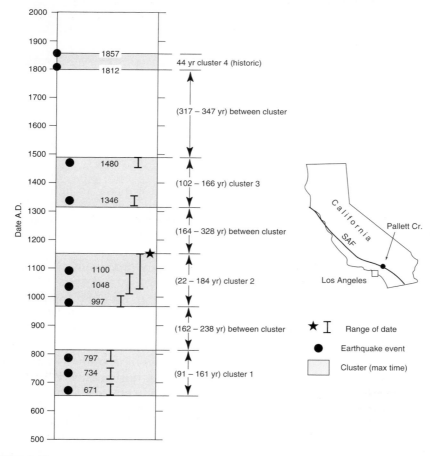

FIGURE 8.23

Graph showing chronology of two historical and eight prehistoric earthquakes on the San Andreas fault at Pallett Creek.

(Data from Sieh et al., 1989 [22])

FIGURE 8.24
Part of a trench wall crossing the San Andreas fault zone at Pallett Creek showing a 30 cm offset. Notice that the strata above the 22 cm scale (central part of photograph) are not offset. The offset sedimentary white beds layers and dark organic layers were produced by an earthquake that occurred approximately 900 yr ago during cluster 2 (see Figure 8.23). (Photograph courtesy of K. Sieh)

dates with their accompanying error bars are plotted, it is apparent that the earthquakes are not evenly distributed through time, but tend to be clustered (Figure 8.23). Most of the earthquakes within clusters are separated by a few decades, but the time between clusters varies from approximately 160 yr to 350 yr. The earthquakes were identified through careful evaluation of natural exposures and fault trenches at Pallett Creek [22]. Figure 8.24 shows part of the trench wall and the offset of organic-rich sedimentary layers that correspond to an earthquake that occurred approximately A.D. 1100.

Of course the question is: When will the next big earthquake occur on the San Andreas fault northeast of Los Angeles? Studies using conditional probability suggest a 30% likelihood for the 30-yr period from 1988 to the year 2018 (see Figure 1.17). This seems reasonable in light of the history of clustering of events on the fault. Assuming that the two historic events are indeed a cluster, and that clusters can be separated by approximately 160 to 350 yr, then the next large earthquake could be expected between the years 2017 and 2207. Thus, a 30% probability for an event by approximately the year 2017 seems reasonable. On the other hand, the data set at Pallett Creek, although the most complete record for any fault in the world, is still a study in small numbers. For example, during cluster 3, the time period between events could have been as long as 166 yr. Thus we might argue that another earthquake in cluster 4 could happen as late as the year 2023 without violating the history at Pallett Creek. However, this seems unlikely given the fact that almost all the time intervals during clusters are closer to 40 to 50 yr.

CONDITIONAL PROBABILITIES FOR FUTURE EARTHQUAKES

Geologists in California have estimated conditional probabilities for major earthquakes along segments of the San Andreas fault for the 30-yr period from 1988 to 2018 (Figure 1.17). These probabilities are based on a synthesis of historical records and geologic evaluation of prehistoric earthquakes [23]. Results suggest that the Parkfield segment is almost certain to rupture by the year 2018. This estimate is based in part on the observation that moderate to large earthquakes have occurred on the Parkfield segment in the years 1857, 1881, 1902, 1922, 1934, and 1966, or on average every 21 to 22 yr. The southern segment of the fault in the Coachella Valley has been assigned a probability of approximately 40% to produce a major earthquake by the year 2018. This study, compiled in early 1989, assigned a probability of about 30% for a major event on the San Andreas fault in the Santa Cruz Mountains where the $M = 7.1$ Loma Prieta earthquake occurred on October 17, 1989. The earthquake supports the validity of the conditional probability approach. On the other hand, the 1992 $M = 7.5$ Landers earthquake occurred east of the San Bernardino Mountains. That event caused ground rupture over nearly 100 km, with maximum right-lateral displacement of about 6 m. Because the displacement is almost entirely pure strike-slip, the fault that ruptured is probably part of the right-lateral plate boundary system between the Pacific and North American plates. Although in retrospect there is evidence for moderate to large earthquakes in the past on the fault that ruptured in 1992, it was not generally recognized that the next right-lateral earthquake in southern California would be so far to the east of the San Andreas fault. There is speculation now that the San Andreas fault, and thus the boundary between the North American and Pacific plates, is either wider than we thought or is in the process of migrating eastward. Of course this does not mean that large earthquakes will not continue to occur on the main San Andreas fault. The conditional probabilities shown on Figure 1.17 still remain our best estimate of the probabilities of large earthquakes occurring in the next few decades on the San Andreas fault.

EARTHQUAKE PREDICTION

Short-term prediction, or forecast, of earthquakes is an area of serious research. The term **forecast** is preferred by some scientists because, like forecasting the weather, an earthquake forecast specifies a time period for the event to occur and a probability of occurrence. The Japanese made the first attempts at earthquake prediction, with some success, using the frequency of microearthquakes, repetitive geodetic surveys, water-tube tiltmeters, and geomagnetic observations. They found that earthquakes in the areas they studied were nearly always accompanied by swarms of microearthquakes that occurred several months before the major shocks. Furthermore, ground tilt correlated strongly with earthquake activity. Anomalous magnetic fluctuations were also reported [24].

Chinese scientists made the first successful prediction of a major earthquake in 1975. The $M = 7.3$ Haicheng earthquake of February 4, 1975, destroyed or damaged about 90% of structures in a city of 9000 people. The short-term prediction of that event was based primarily on a series of foreshocks that began four days prior to the main event. On February 1 and 2, there were several shocks with a $M < 1$. On February 3, less than 24 hr before the main shock, a foreshock of $M = 2.4$ occurred, and in the next 17 hr, eight shocks with $M > 3$ occurred. Then, as suddenly as it began, the foreshock activity became relatively quiet for 6 hr until the main earthquake struck [25, 26]. The lives of thousands of people were saved by massive evacuation from potentially unsafe housing just before the earthquake.

Unfortunately, foreshocks do not always precede large earthquakes. In 1976, a catastrophic earthquake—one of the deadliest in recorded history—struck near the mining town of Tangshan, China, killing several hundred thousand people. There were no foreshocks!

Optimistic scientists around the world today believe that eventually we will be able to make consistent, long-range forecasts (tens to a few thousand years), medium-range predictions (a few years to a few months), and short-range predictions (a few days or hours) for the general locations and magnitudes of large, damaging earthquakes. Unfortunately, earthquake prediction is still a complex problem and it will probably be many years before dependable short-range prediction is possible. Such predictions most likely will be based upon precursory phenomena such as:

- Deformation of the ground surface
- Seismic gaps along faults
- Patterns and frequency of earthquakes
- Changes in electrical resistivity of the Earth
- Changes in the amount of dissolved radioactive gases (radon) in groundwater
- And, perhaps, anomalous behavior of animals

Preseismic Uplift and Subsidence. Rates of uplift and subsidence, especially when rapid or anomalous, may be significant in predicting earthquakes. For example, for more than ten years before the 1964 earthquake near Niigata, Japan ($M = 7.5$), there was broad uplift of the Earth's crust of several centimeters (Figure 8.25) [27]. Similarly, broad, slow uplift of several centimeters occurred over a five-year period prior to the 1983 ($M = 7.7$) Sea of Japan earthquake (Figure 8.25 inset map). The mechanism responsible for the uplift is thought to be deep, stable fault slip prior to the earthquakes [28]. Less well understood, but possibly important observations include pre-instrument measurements of 1 to 2 m of uplift preceding large Japanese earthquakes in 1793, 1802, 1872, and 1927. The uplift was recognized by sudden oceanward shifts of the coastline of as much as several hundred meters. For example, on the morning of the 1802 earthquake (about 10 A.M.), the sea suddenly withdrew about 300 m from a harbor in

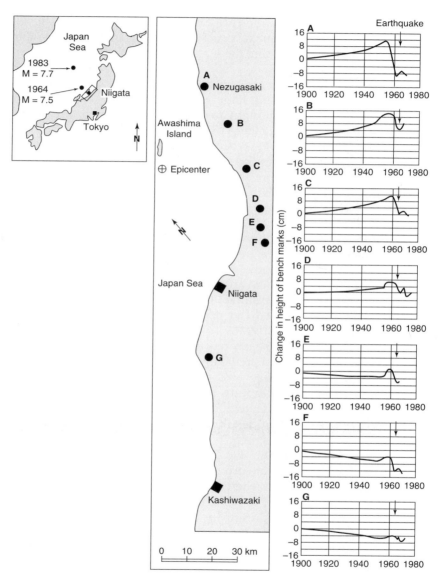

FIGURE 8.25

Anomalous uplift that occurred during the years before the $M = 7.5$ earthquake that struck Niigata, Japan, in 1964. Uplift was measured by plotting changes in the bench marks (points of known elevation) with time. The black dots are the locations of bench marks, and the graphs show the uplift at those locations. The data suggest that there was both uplift and subsidence for several decades until the mid-1950s, when an episode of rapid uplift occurred at all stations. This activity stabilized by about 1960 and was followed by several years of subsidence prior to the 1964 earthquake.

(From Press, 1975. *Earthquake Prediction.* © May, 1975 by Scientific American, Inc. All rights reserved.)

response to a seismic (preseismic) uplift of about 1 m. Four hours later, at 2 P.M., the earthquake struck, destroying many houses and uplifting the land another meter, causing the sea to withdraw farther [28].

Seismic Gaps. Seismic gaps are defined as areas along active fault zones that are capable of producing large earthquakes but have not produced one recently. These areas are thought to store tectonic strain and thus are candidates for future large earthquakes [29].

Seismic gaps have been useful in medium-range earthquake prediction. At least ten large, plate-boundary earthquakes have been successfully forecast since 1965, including one in Alaska, three in Mexico, one in South America, and three in Japan. In the United States, seismic gaps are currently recognized along the San Andreas fault near Fort Tejon, California, where the fault last ruptured in 1857, and in the Coachella Valley, a segment of the fault that has not produced a great earthquake for several hundred years. Both sections are likely candidates to produce a great earthquake in the next few decades [29, 30].

As earth scientists examine patterns of seismicity, two ideas are emerging. First, sometimes there are reductions in small to moderate earthquakes prior to a larger event. For example, prior to the 1978 ($M = 7.8$) Oaxaca, Mexico, earthquake there was a 10-yr period (1963 to 1973) of relatively high seismicity of earthquakes (mostly $M = 3$ to $M = 6.5$), followed by a quiet period of 5 yr. Renewed activity began ten months before the $M = 7.8$ event, which was the basis of a successful prediction [28]. Second, small earthquakes may tend to ring an area where a larger event will eventually occur. Such a ring (or "donut") was noticed in the 16 months prior to the 1983, $M = 6.2$ earthquake in Coalinga, California. Unfortunately, hindsight is clearer than foresight, and this ring was not noticed or identified until after the event.

Electrical Resistivity. Electrical resistivity is a measure of a material's ability to conduct an electrical current. Conductors, such as copper and aluminum, have a very low resistance, whereas as other materials such as quartz (an insulator) have a very high resistance to electrical current. In general, the Earth is a good conductor of electricity, but its resistivity varies with the amount of groundwater present and many other factors. If a current is fed into the Earth between two points separated by several kilometers, changes in the voltages will be noticed if any change in electrical resistivity takes place in the rocks. Changes before earthquakes have been reported in the United States, Eastern Europe, and China (Figure 8.26) [27]. The mechanisms responsible for the decrease in resistivity are related to physical changes that accompany the earthquake cycle (see Figure 1.23). As rocks dilate, there is an influx of water prior to an earthquake and the electrical resistance decreases. Following the earthquake, fluids drain and resistivity increases.

Amount of Radon. The radioactive gas radon occurs naturally dissolved in water of deep wells. In some cases, the concentration of the gas increases significantly in the hours to days before an earthquake (Figure 8.27). The technique of monitoring radon gas has been studied in Eastern Europe, China, and the United States and appears promising. The increase in radon probably results from precursory cracking and dilation of rocks before an earthquake, allowing more radon to dissolve into the groundwater.

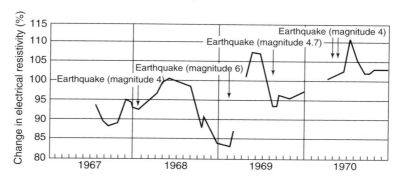

FIGURE 8.26

Changes in electrical resistivity of the Earth's crust before earthquakes were measured in the former USSR, China, and the United States. The data for this graph represent earthquakes monitored in the former USSR between 1967 and 1970. The measurements are made by feeding an electric current into the ground and recording voltage changes a few kilometers away. In general, it has been observed that earthquakes are preceded by a decrease in resistivity.

(From Press, 1975 [27])

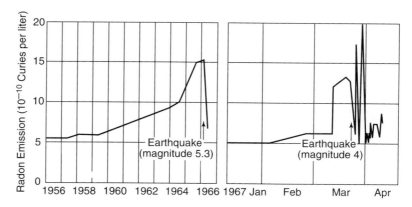

FIGURE 8.27

The amount of radon dissolved in the water of deep wells may increase significantly before an earthquake. The two examples shown here were recorded before earthquakes in the former USSR. The 1966 event (left) had a $M = 5.3$; the 1967 aftershock (right) had a $M = 4$. The technique is used in both the former USSR and China, and is being tried in the United States.

(From Press, 1975 [27])

Anomalous Animal Behavior. Anomalous animal behavior often has been reported prior to large earthquakes; everything from unusual barking of dogs to chickens that refuse to lay eggs, horses or cattle that run in circles, rats that come out of the ground and perch on power lines, and snakes that crawl out of the ground in the winter and freeze. Anomalous behavior reportedly was common before the Haicheng earthquake [25]. Unusual animal behavior also was observed prior to the 1971 San Fernando earthquake. Unfortunately, the significance and reliability of animal behavior is difficult to evaluate. Nevertheless, there has been considerable interest in the topic, and research is ongoing.

Progress Toward Earthquake Prediction

We are still a long way from a working, practical methodology to predict earthquakes reliably. On the other hand, a good deal of data currently is being gathered concerning possible precursor phenomena associated with earthquakes.

Although progress on short-range earthquake prediction has not matched expectations, medium- to long-range forecasting, including hazard evaluation and probabilistic analysis of areas along active faults, has progressed faster than expected [21]. The Borah Peak earthquake of October 28, 1983, in central Idaho has been lauded as a success story for medium-range earthquake-hazard evaluation. Previous evaluation of the Lost River fault suggested that the fault was active [31]. The earthquake, which was about $M = 7$, killed two people and did about $15 million damage, producing fault scarps several meters high and numerous ground fractures along the 36 km rupture zone of the fault. The important fact was that the scarp and faults produced during the earthquake were superimposed on previously existing fault scarps, validating the usefulness of careful mapping of scarps produced by prehistoric earthquakes. The principle is that where the ground has broken before, it may break again!

EARTHQUAKE-HAZARD REDUCTION

The United States is developing a National Earthquake Hazard Reduction Program in cooperation with the United States Geological Survey and other scientists. The major goals of the program [32] are to:

- Develop an understanding of earthquake sources. This involves an understanding of the physical properties and mechanical behavior of faults as well as development of quantitative models of the physics of the earthquake process (see Figure 1.23).
- Determine earthquake potential. This involves characterizing seismically active regions, including determination of the rates of crustal deformation, identification of active faults, determining characteristics of paleoseismicity,

calculating long-term probabilistic forecasts, and, finally, developing methods for intermediate- and short-term prediction of earthquakes.
- Predict effects of earthquakes. This includes gathering data necessary for predicting ground rupture and shaking (see Chapter 1), and the response of buildings and other structures in areas likely to be affected by earthquakes. This goal also involves evaluating the losses associated with earthquake hazards.
- Apply research results. At this level, the program is interested in the transfer of knowledge about earthquake hazards to people, communities, states, and the nation. This knowledge concerns what can be done to better plan for earthquakes and reduce potential losses of life and property.

Earthquakes and Critical Facilities

Critical facilities are those that, if damaged or destroyed, might cause significant to catastrophic loss of life, property damage, or disruption of society. In the urban environment, examples include schools, medical facilities, police stations, and fire stations. Other examples include dams, power plants, and other necessary facilities. Three aspects of the decision-making process concerning critical facilities and earthquake hazard [33] are:

- Evaluation of the hazard
- Evaluation of whether the facility may be designed or modified to accommodate the hazard
- Subjective evaluation of an "acceptable risk"

The first two factors have a strong scientific component, while risk assessment is a public-safety issue, since no facility can be rendered absolutely safe.

In most cases regarding siting of critical facilities, the main scientific challenge is in estimating the activity of a particular fault system. For example, the seismic evaluation for the site of Auburn Dam near Sacramento, California, was very controversial. Millions of dollars were spent and hundreds of meters of fault trenches were excavated during geological studies, and there was still no agreement about the hazard associated with the system of faults in the vicinity of the proposed site. The general area was considered to have a relatively low seismic risk until a $M = 5.7$ earthquake occurred in 1975 near Oroville, about 80 km from the Auburn site. This earthquake rekindled concern and initiated a new round of seismic-risk evaluation, and ultimately led to the virtual abandonment of the site. It was determined from the evaluation that the faults on and near the site would produce only a very small displacement, but this was enough to cause major concern because the concrete dam being planned could not accommodate even such small displacements.

The main problem remains: It is often very difficult to absolutely date prehistoric earthquakes. However, we are making progress in understanding earthquakes, identifying active faults, and estimating the maximum credible earthquake for a particular area.

DEVELOPMENT OF EARTHQUAKE WARNING SYSTEMS

It is technically feasible to develop a warning system that might provide up to a one minute warning to an urban area like Los Angeles prior to the arrival of damaging earthquake waves. This is based on the principle that radio waves travel much faster than seismic waves. The Japanese have had a warning system in place for nearly 20 years that provides warning for their high-speed trains and keeps them from derailing in the event of an earthquake. A proposal for California would be a more sophisticated system of seismometers and transmitters along the San Andreas fault that would first sense motion associated with a large earthquake and then send a warning to the city, which would relay it to critical facilities, schools, and the general population (Figure 8.28). Depending upon where the rupture initiated, the warning

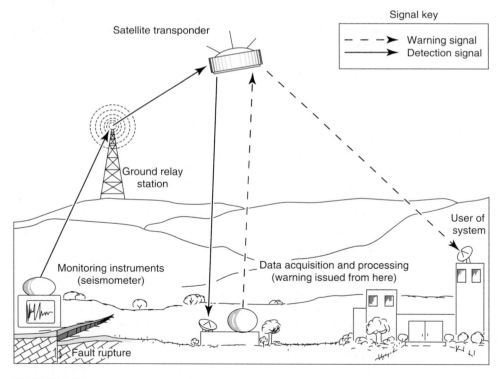

FIGURE 8.28
Idealized diagram showing an earthquake warning system.
(From Holden et al., 1989 [34])

time would vary from 15 seconds to as long as about 1 minute. The warning could provide time for people to shut down machinery and computers and take cover.

A potential problem with this system is the chance of false alarms. Using the Japanese system for comparison, the number of false alarms would be less than 5%. Furthermore, because the length of warning time is so short, some people doubt whether much evasive action could be taken. There also is concern about legal liability resulting from false alarms, warning-system failures, and damage and suffering resulting from actions taken as the result of early warning.

We must emphasize that an earthquake warning system is not a prediction tool, as it *only* provides a warning that an earthquake has already occurred. One study of a potential early-warning system in California concluded that such a system is technically feasible, but the benefits-to-cost ratio makes it economically unfeasible. That is, the study concluded that the potential economic benefits are not sufficient to justify the construction and operating costs. Total cost for the construction was estimated to be about $3 million to $6 million, with annual operating costs of $1.5 million to $2.5 million [34]. For comparison, damages from the 1994 Northridge earthquake with a $M = 6.7$ exceeded $20 billion. Larger, more damaging events can be expected. However, it is difficult to place an economic value on all issues related to public safety, especially as they relate to potential injuries or loss of life. Rejection of technology that could provide warning of approaching, potentially dangerous seismic waves based mainly on economics may be unwise.

ADJUSTMENTS TO EARTHQUAKE ACTIVITY

The mechanisms that trigger earthquakes are still poorly understood, and therefore earthquake prevention and warning systems are not yet viable alternatives. There are, however, reliable protective measures we can take.

- Structural protection, including the construction of large buildings able to accommodate at least moderate shaking. This has been relatively successful in the United States. The 1988 Armenia earthquake ($M = 6.8$) was slightly larger than the 1994 Northridge, California, event ($M = 6.7$), but the loss of life and destruction in Armenia was staggering—at least 45,000 killed (compared with 61 in California) and near-total destruction in some towns near the epicenter. Most buildings in Armenia were constructed of unreinforced concrete and instantly crumbled, crushing or trapping their occupants.
- Land-use planning, or earthquake zonation (also called microzonation), includes the siting of important structures such as schools, hospitals, and police stations in areas away from active faults or sensitive earth materials likely to accentuate seismic shaking.
- Increased insurance and relief measures to help recovery following earthquakes. Following the 1994 Northridge earthquake, total insurance claims were very large, and some insurance companies stopped issuing earthquake insurance. Earthquake zonation could help insurance companies identify high hazard areas and assign premiums accordingly.

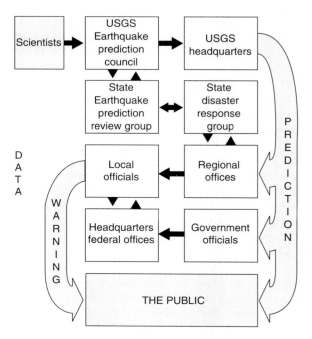

FIGURE 8.29
Proposed information flow for issuing earthquake predictions and warnings.
(From McKelvey, 1976. *United States Geological Survery Circular* 729)

A fourth possible measure is to take little or no action in advance and to pay the consequences when an earthquake occurs. This philosophy is not what we espouse, but it is in fact what we often do.

At the same time that we are adjusting to existing hazards, we are increasing the need for adjustment by creating new hazards where none existed before. When any new building or structure is built on sensitive earth materials or on active faults, we are creating a new problem where none existed before. We know from studying earthquake damage that building on unconsolidated deposits that become unstable (subject to liquefaction) or amplify shaking during an earthquake (see Chapter 1) is more hazardous than building on bedrock. We also know that building astride active faults is unwise.

We hope, eventually, to be able to predict earthquakes. The federal plan for issuing prediction and warning is shown on Figure 8.29. The general flow of information is from scientists to a prediction council for verification. A prediction that a damaging earthquake of a specific magnitude will occur at a particular location over a specified time span then may be issued to state and local officials, who will be responsible for issuing a warning to the public to take defensive action (that has, one hopes, been planned in advance). Potential responses to a

TABLE 8.3
Potential response to an earthquake prediction with given lead time.

Lead Time	Buildings	Contents	Lifelines	Special Structures
3 Days	Evacuate previously identified hazards	Remove selected contents	Deploy emergency materials	Shut down reactors, petroleum products pipelines
30 Days	Inspect and identify potential hazards	Selectively harden (brace and strengthen) contents	Shift hospital patients; alter use of facilities	Draw down reservoirs, remove toxic materials
300 Days	Selectively reinforce		Develop response capability	Replace hazardous storage
3000 Days		Revise building codes and land-use regulations; enforce condemnation and reinforcement		Remove hazardous dams from service

(From Thiel, 1976. *U.S. Geological Survey Circular* 729)

prediction depend upon lead time (Table 8.3), but even a short time (as little as a few days) would be sufficient to mobilize emergency services, shut down important machinery, and evacuate particularly hazardous areas.

SUMMARY

Paleoseismicity is defined as the study of the occurrence, size, timing, and frequency of prehistoric earthquakes. Paleoseismic data may be gathered from fault exposures, faulted landforms, fault scarps, stratigraphic features, and folded deposits, rocks, and geomorphic surfaces.

The concept of fault-zone segmentation is important in evaluation of earthquake hazards. Earthquake segments may be identified in terms of rupture length, determined from historic earthquakes or paleoseismic evaluation. Faults may also be structurally segmented, based upon changes in the surface expression of subsurface fault-zone structure or direct structural evidence. The concept of fault segmentation is particularly important as it relates to long-term earthquake forecasting involving probabilistic assessment of seismic hazard. The concept also is useful in constraining estimates of the maximum earthquake likely to occur along a particular fault and in constraining estimates of seismic ground motion.

Short-term prediction of earthquakes is an area of serious research. Optimistic scientists believe that eventually we will be able to make consistent long-range forecasts, medium-range predictions, and short-range predictions for the general locations and magnitudes of damaging earthquakes. Such predictions will be based upon factors such as deformation of the ground surface, seismic gaps along faults, pattern and frequency of earthquakes, changes in electrical resistivity of the Earth; changes in the amount of dissolved radon gas and groundwater, and, perhaps, anomalous behavior of animals.

The United States, in cooperation with the U.S. Geological Survey and other scientists, is developing programs for earthquake-hazard reduction. Major goals are to develop an understanding of earthquake source, to determine earthquake potential, to predict effects of earthquakes, and to apply research results to minimize the earthquake hazard to people.

REFERENCES CITED

1. Crone, A.J., 1987. Introduction to directions in paleoseismology. In A.J. Crone and E.M. Omdahl (eds.), Directions in Paleoseismology. *U.S. Geological Survey Open-File Report* 87–673.

2. Wallace, R.E., 1987. A perspective of paleoseismology. In A.J. Crone and E.M. Omdahl (eds.), Directions in Paleoseismology. *U.S. Geological Survey Open-File Report* 87–673.

3. Carver, G.A., and R.M. Burke, 1992. Late Cenozoic deformation on the Cascadia subduction zone in the region of the Mendocino triple junction. In *Friends of the Pleistocene, Pacific Cell Guidebook.* Humboldt State University: California.

4. Plafker, G., 1987. Application of marine-terrace data to paleoseismic studies. In A.J. Crone and E.M. Omdahl (eds.), Directions in Paleoseismology. *U.S. Geological Survey Open-File Report* 87–673.

5. Geophysics Study Committee, 1986. Overview and recommendations. In *Active Tectonics.* National Academy Press: Washington, D.C.

6. Wallace, R.E., 1977. Profiles and ages of young fault scarps, north-central Nevada. *Geological Society of America Bulletin,* 88: 1267–1281.

7. Colman, S.M., and K. Watson, 1983. Ages estimated from a diffusion equation model for scarp degradation. *Science,* 221: 263–265.

8. Andrews, D.J., and R.C. Bucknam, 1987. Fitting degradation of shoreline scarps by a nonlinear diffusion model. *Journal of Geophysical Research,* 92: 12,857–12,867.

9. Pierce, K.L., and S.M. Colman, 1986. Effect of height and orientation (microclimate) on geomorphic degradation rates and processes, late-glacial terrace scarps in central Idaho. *Geological Society of America Bulletin,* 97: 869–885.

10. Hanks, T.C., and D.J. Andrews, 1989. Effects of far-field slope on morphological dating of scarplike landforms. *Journal of Geophysical Research,* 94: 565–573.

11. McCalpin, J., 1987. Geologic criteria for recognition of individual paleoseismic events in extensional environments. In A.J. Crone and E.M. Omdahl (eds.), Directions in Paleoseismology. *U.S. Geological Survey Open-File Report* 87–673.

12. Holzer, T.L., and M.N. Clark, 1993. Sand boils without earthquakes. *Geology,* 21: 873–876.

13. Atwater, B.F., 1987. Evidence for great Holocene earthquakes along the outer coast of Washington State. *Science,* 236: 942–944.

14. Allen, C.R., 1968. The tectonic environments of seismically active and inactive areas along the San Andreas fault system. In W.R. Dickinson and A. Grantz (eds.), *Proceedings of the Conference on Geologic Problems of the San Andreas Fault System.* Stanford University Publications, Geological Sciences, 11: 70–82.

15. Schwartz, D.P., and R.H. Sibson, 1989. Introduction to workshop on fault segmen-

tation and controls of rupture initiation and termination. In D.P. Schwartz and R.H. Sibson (eds.), Fault Segmentation and Controls of Rupture Initiation and Termination. *U.S. Geological Survey Open-File Report* 89–315.

16. DePolo, C.M., D.G. Clark, D.B. Slemmons, and W.H. Aymard, 1989. Historical Basin and Range province surface faulting and fault segmentation. In D.P. Schwartz and R.H. Sibson (eds.), Fault Segmentation and Controls of Rupture Initiation and Termination. *U.S. Geological Survey Open-File Report* 89–315.

17. Machette, M.N., S.F. Personius, A.R. Nelson, D.P. Schwartz, and W.R. Lunde, 1989. Segmentation models and Holocene movement history of the Wasatch Fault Zone. In D.P. Schwartz and R.H. Sibson (eds.), Fault Segmentation and Controls of Rupture Initiation and Termination. *U.S. Geological Survey Open-File Report* 89–315.

18. Schwartz, D.P., and K.J. Coppersmith, 1984. Fault behavior and characteristic earthquakes: examples from the Wasatch and San Andreas Fault Zones. *Journal of Geophysical Research*, 89: 5681–5698.

19. Schwartz, D.P., and K.J. Coppersmith, 1986. In *Active Tectonics*. National Academy Press: Washington, D.C.

20. Wallace, R.E., 1970. Earthquake recurrence intervals on the San Andreas fault. *Geological Society of America Bulletin*, 81: 2875–2890.

21. Allen, C.R., 1983. Earthquake prediction. *Geology*, 11: 682.

22. Sieh, K., M. Stuiver, and D. Brillinger, 1989. A more precise chronology of earthquakes produced by the San Andreas fault in southern California. *Journal of Geophysical Research*, 94: 603–623.

23. Heaton, T.H., D.L. Anderson, W.J. Arabasz, R. Buland, W.L. Ellsworth, S.H. Hartzell, T. Lay, and P. Spudich, 1989. National seismic system science plan. *U.S. Geological Survey Circular* 1031.

24. Pakiser, L.C., J.P. Eaton, J.H. Healy, and C.B. Raleigh, 1969. Earthquake prediction and control. *Science*, 166: 1467–1474.

25. Raleigh, B., et al., 1977. Prediction of the Haicheng earthquake. *Transactions of the American Geophysical Union*, 58: 236–272.

26. Simons, R.S., 1977. Earthquake prediction, prevention and San Diego. In A. L. Patrick (ed.), *Geologic Hazards in San Diego*. San Diego Society of Natural History: San Diego, California.

27. Press, F., 1975. Earthquake prediction. *Scientific American*, 232 (May): 14–23.

28. Scholz, C.H., 1990. *The Mechanics of Earthquakes and Faulting*. Cambridge University Press: New York.

29. Rikitakr, T., 1983. *Earthquake Forecasting and Warning*. D. Reidel: London.

30. Hanks, T.C., 1985. The national earthquake hazards reduction program: Scientific status. *U.S. Geological Survey Bulletin* 1659.

31. Hait, M.H., 1978. Holocene faulting, Lost River Range, Idaho. *Geological Society of America Abstracts with Programs*, 10(5): 217.

32. Page, R.A., D.M. Boore, R.C. Bucknam, and W.R. Thatcher, 1992. Goals, opportunities, and priorities for the USGS Earthquake Hazards Reduction Program. *U.S. Geological Survey Circular* 1079.

33. Cluff, L.S., 1983. The impact of tectonics on the siting of critical facilities. *EOS: Transactions, American Geophysical Union*, 64: 860.

34. Holden, R., R. Lee, and M. Reichle, 1989. Technical and economic feasibility of an earthquake warning system in California. *California Division of Mines and Geology Special Publication* 101.

9
Mountain Building

INTRODUCTION

Landscape Scale

If the Earth were as smooth as a cue ball, it would not be a very scenic place, and there would be no such science as geomorphology. The Earth's surface is varied and interesting because it has **relief**—from individual hills and valleys to great mountain chains. The difference between a hill and a mountain is a difference in **scale,** both in terms of the heights of the landforms and their lateral dimensions. In any science, scale is everything! We imagine that atoms look like solar systems, but that is a metaphor, not physical similarity. Hills are different from mountain chains because they form by different processes, have different structures, and persist over different time scales.

The importance of landscape scale has been recognized for a long time, but geomorphologists have not yet agreed upon a single classification system. Joseph Le Conte was an eminent geologist during the latter half of the nineteenth century and the early twentieth century. Le Conte classified relief as either *greater* or *lesser* [1]. According to this classification, greater relief includes the continents, ocean basins, and great mountain chains and is created by processes occurring in the Earth's interior. Lesser relief would include all smaller landforms shaped principally by erosion. Le Conte lived well before the revelation of plate-tectonic theory and did not recognize that the internal processes of the Earth are responsible for many small landforms. Furthermore, as this chapter will discuss, we are increasingly aware that erosional processes play a part in the development of regional topography in addition to its role in sculpting local landforms. Classifications of geomorphic scale since the discovery of tectonics (for example, Table 9.1) tend to focus more on quantification and description than the older schemes [2].

This chapter examines the development of mountain ranges, which is landscape at the scale of Orders 2 and 3 on Table 9.1. The previous eight chapters of this book focused on local features caused directly or indirectly by tectonic activity. Mountain ranges are formed by individual uplift events and gradual uplift processes integrated over the eons of geological time. Since geologists began studying mountains, several refinements in the terminology have been proposed, generally based on scale. Le Conte used the following terms, from most local scale to most regional: *mountain peak, mountain ridge, mountain range, mountain system,* and *cordillera* (continental-scale mountainous topography). We

TABLE 9.1

A classification of landforms by scale.

Order	Lateral dimensions (km)		Examples	Le Conte classification
1	>10,000 km		continents, ocean basins	
2	1000–10,000 km	tectonic units	physiographic provinces, mountain belts	Greater Relief
3	100–1000 km		sedimentary basins, mountain ranges	
4	10–100 km		volcanoes, fault-block mountains, troughs	
5	1–10 km	erosional-depositional units	deltas, piedmonts, major valleys	Lesser Relief
6	100 m –1 km		floodplains, alluvial fans, moraines	
7	10–100 m		terraces, sand dunes, gullies	
8	1–10 m	geomorphic process units	hillslopes, sections of stream channels	
9	10 cm –1 m		river bars, pools and riffles, beach cusps	
10	1–10 cm		fluvial or eolian ripples, glacial striations	

(After Baker, 1986 [2]; Le Conte, 1909 [1])

prefer a more modern, less burdensome vocabulary that includes only three different scales: *mountain peak* (an individual peak, such as a single volcanic cone), *mountain range* (a linear trend of mountain peaks), and *mountain chain* or *belt* (a series of mountain ranges related by a common origin and/or structure) [1].

Models of Landscape Development

There may be as many different theories of landscape development as there have been geologists who have studied the subject. But three of those theorists have shaped the modern understanding of landscape more than most others: William Morris Davis, Walther Penck, and John T. Hack. Davis' "Geographical Cycle" was introduced in Chapter 2 and illustrated in Figure 2.1. In brief, Davis proposed that landscapes are subject to brief periods of tectonism and mountain building, followed by long periods of erosion. In this cycle, a landscape goes from a stage of "youth," characterized by deep incision, to a "mature" stage, with a fully developed drainage network and moderate relief, to "old age" [3]. Landscapes in old age are characterized by **peneplains,** which are regional surfaces of low relief, produced by long-continued erosion in the Davis model.

A problem with Davis' model is that his view of episodic tectonism runs contrary to the geological *principle of uniformitarianism*—the principle that modern processes have operated throughout the geological past. Penck, a contemporary of Davis, sidestepped that criticism by proposing a model in which topography is a function of continuous uplift, at rates that accelerate and decelerate over time [4]. Modern theories of landscape development have moved even further toward processes that operate continuously. Hack proposed the existence of "steady-state" landscapes, in which constructive and destructive processes may be balanced in a state of "dynamic equilibrium" [5].

HYPSOMETRY: THE SHAPE OF THE EARTH

The concept of hypsometry—the distribution of different elevations across a landscape—was introduced in Chapter 4. Graphs of elevation-frequency distribution (Figure 9.1) can illustrate the differences between landscapes, such as the three types discussed above (youthful, mature, and old-age landscapes). Hypsometry can be evaluated over small areas, such as a single drainage basin (see Chapter 4), or over large areas, including the entire planet (Figure 9.2). The total elevation-frequency distribution of the Earth reveals much about the structure of the planet and the processes operating in the interior and on the surface.

The mean elevation of the Earth is 2.3 km below sea level [6], but the distribution is strongly bimodal (Figure 9.2). One mode is between about 3.0 km and 5.0 km below sea level, and the other is a sharp spike between about 0.2 km below and 0.5 km above sea level. This bimodal distribution represents the

FIGURE 9.1
The elevation-frequency distribution for three different stages of landscape development. An uplifted landscape is mountainous terrain, with high relief and high absolute elevation. A rejuvenated landscape has a broad upland surface, deeply incised by canyons. An eroded landscape is of low relief and low absolute elevation.

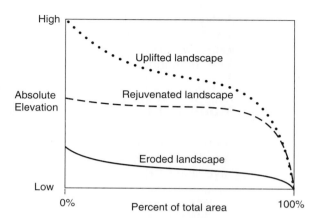

profound differences between oceanic crust and continental crust (see Chapter 1). Oceanic crust is thin (<10 km), dense (3000–3100 kg/m³), and composed of basaltic rock [7]. In contrast, continental crust is thick (35–40 km, on average), predominantly granitic, and relatively buoyant (2700–2800 kg/m³), and it therefore stands higher than the ocean basins. The abrupt transition from continental crust to oceanic crust, at the edge of the continental shelves at depths of 100–200 m below sea level, contributes to the distinct separation of elevations in Figure 9.2.

The Earth's surface also has been profoundly shaped by the presence of its hydrosphere (oceans, lakes, rivers, glaciers, groundwater, etc.). The oceans weigh down upon oceanic crust and accentuate the contrasts in Figure 9.2. Furthermore, the concentration of the Earth's surface elevation near sea level is a

FIGURE 9.2
The elevation-frequency distribution of the Earth. The curve indicates the percent of area for each 200-m interval. Data are averages from 1° by 1° areas; therefore the range indicated is less than the total range of elevation on Earth. (After Head et al., 1981 [8])

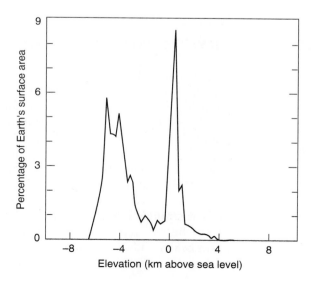

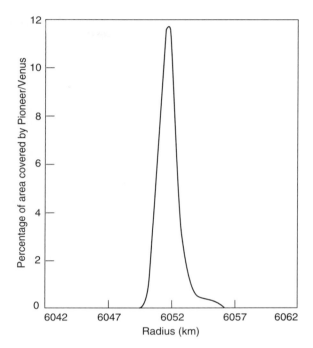

FIGURE 9.3
The elevation-frequency distribution of Venus. Data is radar altimetry from the *Pioneer/Venus* spacecraft, with a resolution between 100 and 200 km.
(After Head et al., 1981 [8])

result of the tireless force of running water—areas above sea level erode to the coastal base level, and the resulting detritus fills shallow marine basins.

Lest anyone become complacent about the peculiar topography of the Earth, compare Figure 9.2 with Figure 9.3, which illustrates the hypsometry of the planet Venus [8]. Venus has only one type of crust—basaltic—and hence a unimodal height distribution. With an average surface temperature of about 200°C and atmospheric pressure up to 90 times that of Earth [9], Venus is shaped by processes completely alien to us. At these surface conditions, Venus has only a thin, brittle crust over the ductile mantle below, that forms a lithosphere without distinct plates like Earth's, but which rather deforms by countless small-scale rips and tears across zones up to several hundred kilometers wide [10]. However, in spite of profound differences, the tectonic processes on both planets have led to the formation of mountain ranges (the extremely high elevations seen as small "tails" on both Figure 9.2 and Figure 9.3). For example, on Venus, a broad zone of crustal convergence has uplifted a tall mountain chain (Freyja Montes) and a high-standing plateau (Lakshmi Planum) (Figure 9.4) that are directly comparable to the Himalaya and Tibetan Plateau [11, 12].

DRIVING MECHANISMS FOR OROGENY

The principal driving mechanism for **orogeny** (large-scale regional mountain building) is plate tectonics. Recall that the interiors of the lithospheric plates

FIGURE 9.4
Radar image of central Freyja Montes region of Venus from *Magellan* spacecraft. The region shown is interpreted to be a collision zone, very similar to the Himalaya of Earth. (Image courtesy of Jet Propulsion Laboratories)

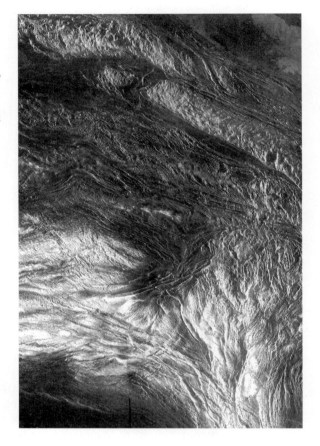

are relatively stable; the edges are where the action is. Plate boundaries are either divergent, convergent, or transform. Each type of boundary is characterized by high-standing topography. Divergent plate boundaries (mid-ocean ridges and continental rifts) stand higher than stable plate interiors because of heating of the crust. Transform plate boundaries also may be the site of significant uplift, but only where strike-slip motion includes a component of convergence. However, true orogeny, uplift of the largest areas to the greatest elevations, occurs at convergent plate boundaries, at subduction zones and continental collisions.

Convergence between oceanic lithosphere and continental or oceanic lithosphere (subduction) causes compression and thickening of the overriding plate. Partial melting of the downgoing plate also generates magma, which is added to the area above. The classic example of a mountain chain formed by these processes is the Andes of South America. The modern Andes

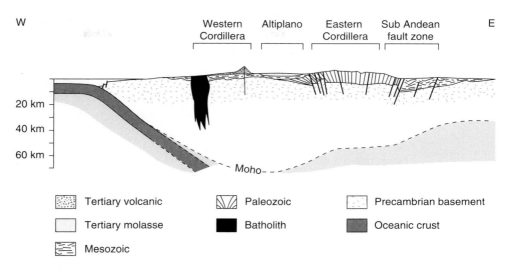

FIGURE 9.5
Diagrammatic cross section through the Peruvian Andes.
(After Cobbing and Pitcher, 1972. *Nature Physical Science*, 240: 51–53)

began forming in the early Jurassic (around 200 Ma), when subduction
began along the western margin of Pangea (the supercontinent of which
South America was a part at the time) [13]. In the southern Andes, as much
as 240 km of horizontal shortening has accompanied subduction [14]. The
modern Andes reflect this compression, as well as the history of volcanism
and erosion (Figure 9.5). The Western Cordillera rises 5000 m from the
Pacific and the narrow coastal zone. The Eastern Cordillera rises even further,
to 6000 m, and is separated from the western ranges by the Altiplano ("high
plain") in the Peruvian and Bolivian Andes. East of the Eastern Cordillera is
the Subandean zone, foothills of the Andes up to 2000 m high that constitute
a Pliocene-age belt of folding and thrusting [15]. Variations in the morphology of the Andes today are related to variations in the geometry of the subducting plate. Where the plate subducts at a low angle (~10°, northern and
central Peru), there is no history of recent volcanism, but there is widespread
seismicity; where the plate subducts at a steeper angle (~30°, southern Peru),
there has been volcanism, and seismicity has been concentrated along the
downgoing plate [16].

Uplift, volcanism, and faulting in subduction-zone mountain ranges can
occur over long periods of geological time; indeed, they will occur as long as the
plates converge and oceanic lithosphere is subducted. However, a different kind
of mountain building occurs when plate convergence can no longer be accommodated by subduction. For example, 80 Ma ago, the Indian continent and Asia

were separated by 4000 to 5000 km of ocean (the Tethys Sea) (Figure 9.6). Oceanic crust of the Indian plate was subducted beneath the southern margin of the Eurasian plate, which formed a volcanic mountain chain along the Asian coast similar to the modern Andes. A belt of granite 2500 km long is the eroded remnant of that mountain chain today. At this time, India was moving northward, narrowing the Tethys Sea at a rate of 15–20 cm/yr [17]. By 40–50 Ma, the ocean was closed and continental crust of the Indian plate had begun subducting. As discussed in Chapter 1, continental crust is thick and buoyant and will not descend into the underlying mantle. For this reason, when a continent converges upon another continent, tremendous forces are unleashed. In the 40 m.y. since India ran into Asia, the collision has reshaped the face of Asia, uplifted the Himalaya and the Tibetan Plateau, and altered the climate of the region and perhaps the entire world.

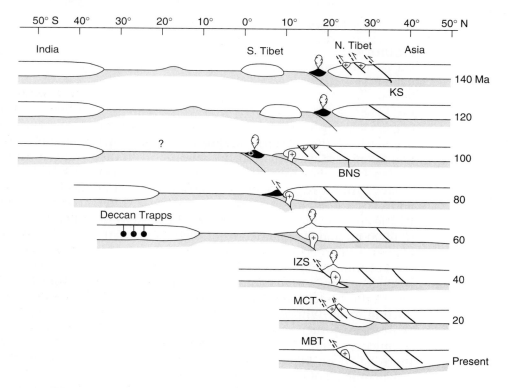

FIGURE 9.6
The closure of the Tethys Sea and collision of India and Asia. Bangong Nujiang suture (BNS); Indus-Zangbo suture (IZS); Kokoxili suture (KS); Main Boundary thrust (MBT); Main Central thrust (MCT).
(From Allègre et al., 1984. *Nature*, 307: 17–22)

A number of continental collisions have occurred in geological history. Such a collision was the culminating event that uplifted the Appalachian Mountains in the Late Paleozoic, around 300 Ma, when North America collided with Africa [18]. In fact, it has been suggested that plate tectonics are fundamentally cyclic, and collision events and rifting will follow one another so long as tectonics operate [19]. In the rocks of ancient collision zones, we have the skeletal records of the processes that occur, but in the Himalaya and Tibet, we have the privilege of watching continental collision in action.

After the Indian continent ran into Asia, the rate of convergence slowed to about 5 cm/yr [17], but that rate has been maintained to the present—a total of over 2000 km of violent compression. In the process, India has been partially thrust beneath Asia [20]. Continued convergence between the two continents appears to have been accommodated by three mechanisms:

1. movement on regional thrust faults
2. continental extrusion
3. uplift

Several hundred kilometers of motion is confirmed on several crustal-scale thrust faults (the Himalayan Frontal thrust, the Main Boundary thrust, and the Main Central thrust). A large portion of the convergence was accommodated by lateral extrusion of portions of Asia along regional strike-slip fault systems (Figure 9.7) [21]. The collision caused at least two major pulses of uplift, one between 21 and 17 Ma and the other between 11 and 7 Ma [22].

Uplift of the Himalaya and the Tibetan Plateau has had several dramatic effects: topography that cross-cuts the modern rivers, voluminous discharge of sediment, and initiation of the monsoonal weather system. The major rivers of the region are **antecedent,** meaning that they predate the collision, and the Himalaya and Tibet have been lifted up beneath the rivers. These rivers now flow through deep gorges—for example, the Indus River flows less than 21 km in distance from Nanga Parbat peak, but over 7000 m lower in elevation [22]! Another effect of uplift has been the erosion and discharge of tremendous volumes of sediment, much of which has accumulated in the vast submarine fans at the outlets of the Indus and Ganges Rivers. Uplift of so large an area has caused region-wide climate changes that have intensified erosion of the mountain range. About 8 Ma, uplift initiated, or at least dramatically strengthened, the seasonal storm system known as the Asian **monsoon** that now shapes life in southern and Southeast Asia [21]. In fact, it has been suggested that uplift of the Himalaya and Tibet altered the global climate and intensified erosion worldwide [23], although there are alternative explanations that will be discussed later in this chapter (see The Problem with Rates).

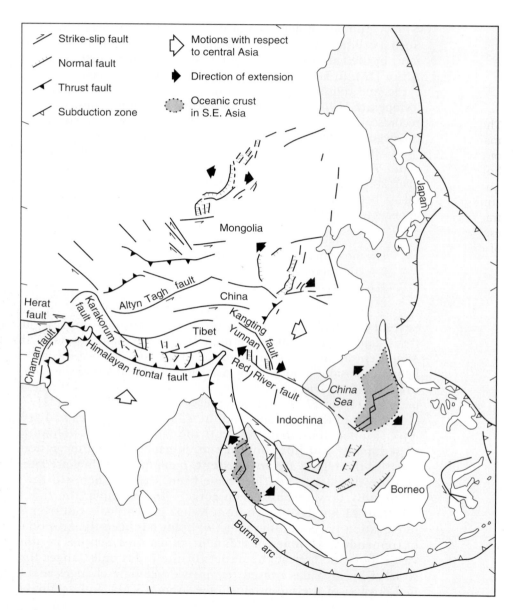

FIGURE 9.7
Extrusion of eastern and southeastern Asia caused by the India-Asia collision. India is interpreted to have penetrated into Asia, causing widespread strike-slip faulting and regional reshaping of the Asian continent.

(From Tapponier et al., 1982. *Geology,* 10: 611–616)

SPECIAL SECTION

ALLUVIAL FANS AND TECTONIC ACTIVITY AT MOUNTAIN FRONTS

The geological forces that lift mountains up may act over large regions, but deformation of the surface often is manifested at discrete boundaries or zones of activity. As earthquakes are concentrated on discrete fault zones, mountain building can be concentrated on distinct **mountain fronts** (a steep escarpment that may reflect long-term uplift of one side of the front relative to the other side; Figure 9.8). The geomorphology of mountain fronts reveals a great deal about the tectonic activity occurring there (see Mountain Front Sinuosity Index in Chapter 4). For example, mountain fronts in arid and semiarid regions are characterized by **alluvial fans** (see Figure 9.8). An alluvial fan may be thought of as the end point of an erosional-depositional system in which sediment

FIGURE 9.8
Front of the White Mountains, California. A single escarpment, about 2500 m high, separates the mountain range from the adjacent valley. The prominent alluvial fan forms where a stream exits the confines of a canyon within the range. Alluvial fans are a common feature at mountain fronts in arid and semiarid settings.

eroded from a mountain source area is transported to the mountain front. At the front, it is deposited as a fan-shaped body (segment of a cone) of fluvial and/or debris-flow deposits [24]. The connecting link between the erosional and depositional parts of the system is the stream. The morphology of an alluvial fan is a function of variables including size of drainage basin contributing sediment to the fan, geology of the source area, relief of the source area, climate and vegetation of the source area, and tectonic activity. Radial profiles of alluvial fans are generally concave but often contain significant breaks that mark boundaries between fan segments. In fact, most alluvial fans are segmented, and younger fan segments may be identified from older segments based on relative soil-profile development, weathering of surficial clasts, erosion of the surface, and development of desert varnish [24, 25].

The morphology of a segmented alluvial fan may be used as an indicator of active tectonics, because the fan forms may reflect varying rates of tectonic processes such as faulting, uplift, tilting, and folding along and adjacent to the mountain front. In the simplest case, if the rate of uplift along a mountain front is high relative to the rate of downcutting of the stream channel in the mountain and to deposition on the fan, then deposition tends to occur in the fanhead area. The youngest fan segment is found near the apex of the fan. Such mountain fronts would also tend to have low values of mountain-front indices S_{mf} and V_f (see Chapter 4), indicating relatively high rates of tectonic activity. On the other hand, if the rate of uplift of the mountain front is less than or equal to the rate of down-cutting of the stream in the mountain source area, then fanhead incision occurs, and deposition is shifted down-fan. As a result, younger fan segments are located well away from the mountain front (Figure 9.9 illustrates these two conditions). Such

mountain fronts would have relatively high value of S_{mf} and V_f, suggesting relatively low rates of mountain-front activity. The mountain block is wearing down with time; the front is more eroded and sinuous, and mountain-front valleys are wider. The situation gets more complex if tilting of the alluvial fan occurs. For example, if an alluvial fan is tilted beyond a threshold slope, then entrenchment may occur and deposition will form a new fan segment away from the mountain front. For example, Hook, working on alluvial fans in Death Valley [26], found that tilting produced segmented fans. Alluvial fans on the east side of the valley are tilted down to the east and this shifted the locus of fan deposition down-fan. That is, fanhead incision has occurred, and younger fan segments are located well away from the mountain front and fan apex. On the east side of Death Valley, the basin is being downdropped relative to the mountain along normal faults, and segments of alluvial fans are often (but not always) located near the apex area at the mountain front. At any rate, the important aspect of studying alluvial fans is that they are sensitive to active tectonics and may provide important information concerning recency and style of tectonic movements.

The overall shape of an alluvial fan also can reveal the pattern of tectonic activity at and near a mountain front. Because alluvial fans are conical in shape, topographic contours across simple fans are approximately circular (the intersection of a cone with a horizontal plane is a circle). Where alluvial fans are not simple, however, and have undergone tectonic tilt, contour lines across the fans form segments of *ellipses*, not circles (the intersection of a *tilted* cone and a horizontal plane is an *ellipse*; Figure 9.10) [27, 28]. Where conditions are appropriate, the amount of tilt can be calculated by fitting ideal ellipses to the contours across an alluvial fan and measuring the length of the long axes *(a)* and the short

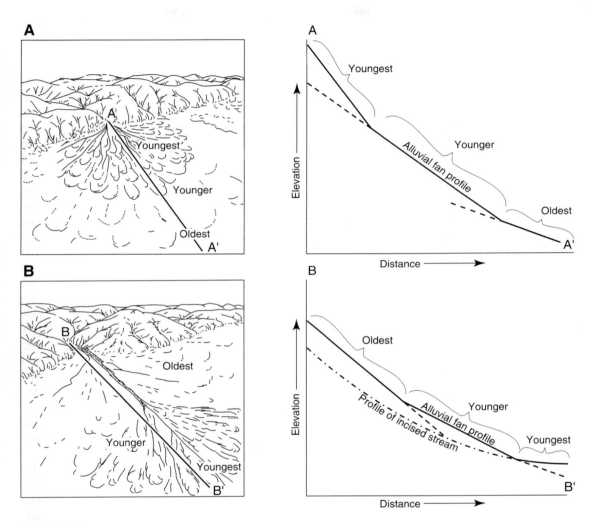

FIGURE 9.9
Alluvial-fan morphology: (A) Fan segments are adjacent to mountain front and are associated with active uplift. (B) Youngest fan segments are away from the mountain front and are associated with erosion of the mountain block rather than uplift.
(After Bull, 1977 [24])

axes *(b)* of the ellipses [29]. It turns out that the amount of tilt *(β)* equals:

$$\beta = \arccos(((b/a)^2 \sin^2 \alpha + \cos^2 \alpha)^{0.5}) \quad (9.1)$$

where α is the slope of the fan along the short axes of the ellipses (see Figure 9.10) [29].

Topographic contours across alluvial fans also may indicate shifts in tectonic activity from one mountain front to another. For example, the northern margin of the San Emigdio Mountains in California is characterized not by one single mountain front, but by several parallel escarpments [30]. Contour

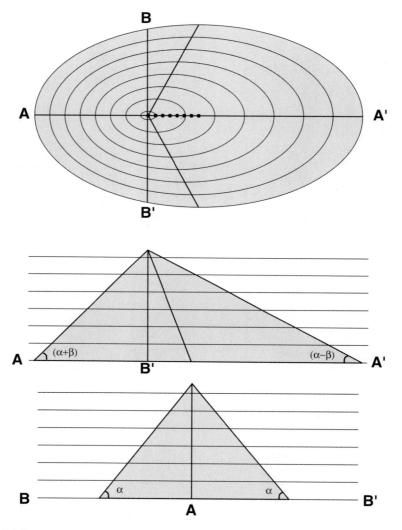

FIGURE 9.10
Geometric model of a tilted cone, viewed from above (top diagram) and in cross section (middle and lower diagrams). Fine lines represent topographic contours (lines of equal elevation across the surface of the cone). Note that the contour lines form ellipses in map view. A tilted alluvial fan consists of some fraction of the total cone shape, typically up to 180° of the total.

lines across alluvial fans there are circular, but the centers of the circles are located along an old mountain front within the range, not at the active mountain front. The conclusion here is that mountain building and alluvial-fan formation occurred at the old front before activity shifted away from the range to its present position.

STRUCTURAL SUPPORT OF MOUNTAIN RANGES

Tremendous forces are required not only to build a mountain range, but also to maintain one. Like a waiter trying to carry a great load of dishes on his tray, at some point the question of how to add another plate to the stack becomes less important than how to keep the entire pile from collapsing to the floor. Indeed, the Tibetan Plateau is now cut by north-south–trending normal faults, indicating that the landmass may have begun to slide to the east and west under its own weight [20].

The primary mechanism of mountain-range support is **isostasy.** Remember that both oceanic crust (density of about 3000 kg/m³) and continental crust (density of about 2700 kg/m³) are less dense than the mantle (density of about 3300 kg/m³ at the base of the crust). For most purposes, the buoyant crust can be thought of as floating upon the mantle below. According to the principle of isostasy, the weight of the crust is borne by the fluid-like properties of the mantle. Two different hypotheses for isostatic support of mountains were proposed in 1855, one by George Biddell Airy, Astronomer Royal of England, and the other by John Henry Pratt, Archdeacon of Calcutta.

According to Airy, any high-standing land mass at the surface is supported by a deep root at the base of the crust (Figure 9.11A). To understand this, visualize two icebergs floating in the ocean, one standing twice as far above the water as the other; we know that the larger iceberg also has correspondingly more ice beneath the surface. The density of ice is about nine-tenths that of water, and thus an iceberg must displace nine kilograms of water to support one kilogram of ice above the waterline. In contrast, according to the Pratt hypothesis, the crust is generally of uniform thickness, but is characterized by regional variations in density. According to this hypothesis, high mountainous topography can be supported by low-density material beneath the mountains (Figure 9.11B). To give an analogy, a block of styrofoam floats much higher than a block of dense wood. We now know that (as is usually the case) there are elements of truth in both hypotheses. The crust beneath mountains is indeed much thicker than in low-lying areas, but it is also true that density differences, such as those caused by heating of the crust, do cause regional uplift in some areas [31].

A third, nonisostatic mechanism exists to support loads upon the surface—**flexural support.** When a skyscraper is built, it does not slowly sink beneath the surface until nine-tenths of it is below ground level, as pure isostasy might suggest. Skyscrapers can be built because the Earth's lithosphere has *strength*. Any local load is supported not only by the buoyancy of the crust directly under it, but also by the rigidity of the lithosphere in a large area around it. Another way to say this is that isostatic support is regional—a skyscraper built in Dallas will depress the lithosphere all over Texas, but not by a measurable amount. A geological example of flexural support is the Himalaya, which are neither thick enough (only 55 km thick;

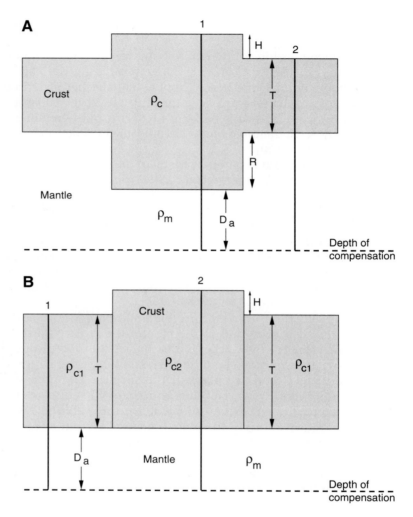

FIGURE 9.11
Two models of isostatic support of mountainous topography. (A) Airy isostasy is based on the premise of deep crustal roots beneath mountains. (B) Pratt isostasy is based on a column made of less-dense material. Variables are explained in the *Special Section* that follows.

they would need to be >80 km for Airy support) nor buoyant enough to support the great mass of the range [32]. Instead, the Himalaya partially rest upon the edge of the Indian plate, flexing India downward and forming the Ganga basin, a deep, sediment-filled trough at the boundary of the two plates [17].

FIGURE 9.12
Support of loads upon the surface (such as mountain ranges) by flexure of the lithosphere. Flexure depends on the strength (rigidity) of the lithosphere. (A) If zero rigidity is assumed, support is purely isostatic. (B) If there is some finite rigidity, the regional flexure occurs. (C) If infinite rigidity is assumed, no isostasy occurs, and the load is carried exclusively by the strength of the lithosphere.

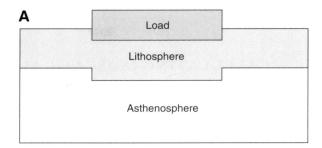

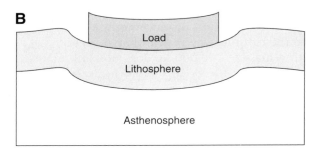

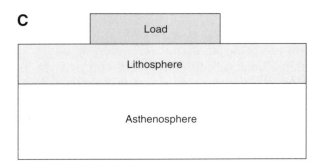

Flexural support of loads on the lithosphere depends on the **wavelength** (the lateral dimensions) of the load. A single skyscraper has a wavelength of no more than 100 m, whereas the Himalaya has a wavelength of hundreds to thousands of kilometers. Flexural support also depends on the strength of the crust. In general terms, thick crust is strong crust; rigidity is measured by the **elastic thickness** (T_e) of the crust, which is the depth to which it will act as a rigid solid. In North America, the elastic thickness of the crust ranges from about 4 km in the Basin and Range province to about 128 km in the cratonic core of the continent [33]. The shorter the wavelength of the load and the greater the rigidity of the crust, the more that flexure will support a load rather than isostasy (Figure 9.12).

SPECIAL SECTION

CALCULATING ISOSTATIC SUPPORT OF MOUNTAINS

Isostasy is based upon **Archimedes principle**—that the weight of a floating solid is supported by the weight of the fluid that it displaces. Archimedes was the ancient Greek scientist who is said to have made his insights about fluid displacements getting into the bathtub. The reason that an iceberg with a mass of 100,000 kg does not sink to the bottom of the ocean is that it displaces exactly 100,000 kg of water. The reason that nine-tenths of the iceberg is below the waterline is that ice is only nine-tenths as dense as water—100,000 kg of water are displaced by 90,000 kg of ice, and the remaining 10,000 kg of the iceberg stand above the surface.

The calculation of isostatic support is based upon the principle that at any given depth within a fluid, called the **depth of compensation,** the pressures generated by overlying material are everywhere equal (see Figure 9.11). This means that the weights of any two columns of material, measured down to the depth of compensation, are equal. Given an area of uniform crustal density (ρ_c) and thickness (T), a mountain block with elevation H above the surface, and an arbitrary distance down to the depth of compensation (Da), we can calculate the mountain-root depth (R) beneath the base of the crust for Airy isostasy (Figure 9.11A). Weight (W) equals mass (volume × density) times gravity (g). For column 1, through the mountain block

$$W_1 = (H\rho_c + T\rho_c + R\rho_c + D\rho_m)\, g \qquad (9.2)$$

The weight of column 2 equals

$$W_2 = (T\rho_c + R\rho_m + D\rho_m)\, g \qquad (9.3)$$

The weights of the two columns equal each other, therefore

$$W_1 = W_2 \qquad (9.4)$$

$$(H\rho_c + T\rho_c + R\rho_c + D\rho_m)\, g \\ = (T\rho_c + R\rho_m + D\rho_m)\, g \qquad (9.5)$$

$$H\rho_c + T\rho_c + R\rho_c + D\rho_m \\ = T\rho_c + R\rho_m + D\rho_m \qquad (9.6)$$

$$H\rho_c + T\rho_c + R\rho_c = T\rho_c + R\rho_m \qquad (9.7)$$

$$H\rho_c + R\rho_c = R\rho_m \qquad (9.8)$$

$$R\rho_m - R\rho_c = H\rho_c \qquad (9.9)$$

$$R = H\rho_c\, /\, (\rho_m - \rho_c) \qquad (9.10)$$

This expression is easily solved to give the depth of the crustal root beneath the Tibetan Plateau, which has an average elevation of 5000 m. We know that the density of continental crust is about 2700 kg/m³ and the density of the mantle is about 3300 kg/m³, therefore

$$R = 5000 \text{ m} \bullet 2700 \text{ kg/m}^3\, / \\ (3300 \text{ kg/m}^3 - 2700 \text{ kg/m}^3) \qquad (9.11)$$

$$R = 5000 \text{ m} \bullet 2700 \text{ kg/m}^3\, / \\ (600 \text{ kg/m}^3) \qquad (9.12)$$

$$R = 22{,}500 \text{ m} \qquad (9.13)$$

In Pratt isostasy (Figure 9.11B), the base of the crust is uniform, and the height of the mountain *(H)* is supported by a lesser density (ρ_{c2}) than the density in the surrounding region (ρ_{c1}). In this problem, we will solve for the value of ρ_{c2}. Once again, we set the weights of the two columns equal to each other:

$$(T\rho_{c1} + D\rho_m)\, g = \\ (H\rho_{c2} + T\rho_{c2} + D\rho_m)\, g \qquad (9.14)$$

$$T\rho_{c1} + D\rho_m = H\rho_{c2} + T\rho_{c2} + D\rho_m \qquad (9.15)$$

$$T\rho_{c1} = H\rho_{c2} + T\rho_{c2} \qquad (9.16)$$

$$T\rho_{c1} = \rho_{c2}(H + T) \qquad (9.17)$$

$$\rho_{c2} = T\rho_{c1} / (H + T). \qquad (9.18)$$

If the Tibetan Plateau were of normal thickness $(T \approx 40 \text{ km})$, equation 9.18 above states that it would require a crustal root with an average density of 2400 kg/m³ to support the additional 5000 m of elevation.

DENUDATION

When discussing mountainous landscapes, uplift processes are, at most, only half of the story. Earthquakes, deformation, and orogeny are impressive, but it is the persistent, sculpting hand of erosion that should be given most of the credit for the rugged topography of the Alps, the Rocky Mountains, and the other ranges of the world. Although uplift temporarily may gain the advantage over periods of tens of millions of years, the geological history of the Earth shows that erosion always has outlasted it and erased the great mountainous landscapes of the past.

It is worth reviewing some of the vocabulary of erosion before discussing the process in greater detail. **Weathering** is the sum of physical and chemical processes that act to break down bedrock at and near the Earth's surface. Weathering is the process that disaggregates and decomposes rock, providing soil, sediment, and solutes to transport processes. **Erosion** is the most general of the terms used; it includes both processes of weathering and of transport. Erosion typically is used to describe both small-scale processes, such as soil creep on a single hillside, to large-scale behavior such as the removal of material from an entire continent over geological time periods. The latter is an example of **denudation,** which specifically refers to regional downwearing, including both weathering and transport processes. This chapter, which focuses primarily on erosion over large areas and over long periods of time, is primarily concerned with denudation.

Rates of denudation are not uniform within a single mountain range, nor from one range to another (Table 9.2). Rates depend on the type of rock being eroded and the geometry of the rock (geological structures). The Yellow River of China and the Colorado River of the United States both drain about equal areas, but the Yellow River drains deposits of loose glacial loess and produces over ten

TABLE 9.2
Denudation rates for various climate and relief conditions.

Climate	Relief	Typical Range for Rate of Denudation (cm/1000 yr)
Glacial	Normal (= ice sheets)	5–20
	Steep (= valley glaciers)	100–500
Polar/montane	Mostly steep	1–100
Temperate maritime	Mostly normal	0.5–10
Temperate continental	Normal	1–10
	Steep	10–20+
Mediterranean	—	1–?
Semiarid	Normal	10–100
Arid	—	1–?
Subtropical	—	1?–100?
Savanna	—	10–50
Rainforest	Normal	1–10
	Steep	10–100
Any climate	Badlands	100–100,000

[Saunders and Young, 1983. *Earth Surface Processes and Landforms,* 8: 473–501].

times as much sediment as the Colorado River [34]. In addition, erosion rates tend to be higher at higher elevations [35, 36]. At higher elevations, local relief tends to be greater, and slopes are steeper, accelerating sediment transport. Precipitation also increases with increasing elevation. Local erosion rates also depend on the particular geomorphic processes at work, which themselves vary with elevation. For example, glacial action is far more erosive than is rainfall. A mountain range that is uplifted to a critical elevation for glacier formation would experience an abrupt increase in erosion.

Another type of critical threshold in mountain uplift and erosion processes occurred during the India-Asia collision. As mentioned earlier, uplift of the Himalaya and Tibetan Plateau is believed to have triggered, or at least dramatically strengthened, the Asian monsoon about 8 Ma. It is also thought by some that the Tibetan Plateau achieved most or all of its current altitude by the same time [21], although geodetic measurements in the Himalaya suggest continuing uplift [37]. The most plausible explanation is that uplift of Tibet has continued since 8 Ma, but the initiation of the monsoon increased erosion rates to near the rate of uplift. As a result, the net elevation of the plateau has not increased much since 8 Ma.

It is important to note that the effects of climate on erosion that are described here are not static in either space or time. Especially in the last 2 m.y., rapid, large-magnitude shifts in climate repeatedly have altered the processes and rates of geomorphic activity across the planet.

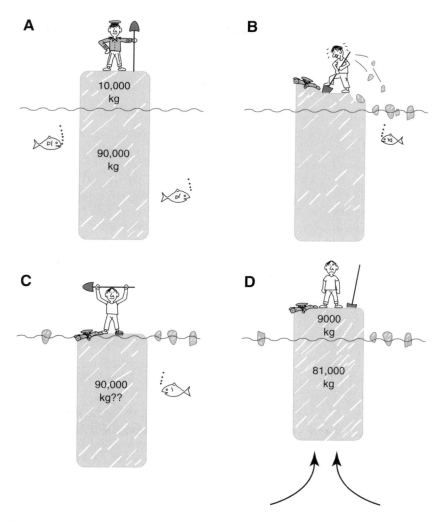

FIGURE 9.13
The principle of isostatic uplift. (A) Admiral Frost, adrift on an iceberg, is uncomfortable so far above the surface of the water. (B) He attempts to remove the 10,000 kg of ice above the waterline. (C) Without isostatic (buoyant) uplift, he would have reached his goal, but in the real world (D), isostasy (buoyancy) always keeps one-tenth of the iceberg above the water.

Isostatic Uplift

On a dynamic Earth, some things are not as they seem. Denudation is the process than wears a landscape down, but as a result of isostasy, it is also a mechanism for uplift [38, 39]. To understand this, recall the example of a 100,000 kg iceberg (Figure 9.13). Imagine that Admiral Frost, intrepid explorer of the Arctic, is adrift on the iceberg and is intent on shoveling away the 10,000 kg of ice above the waterline (Fig 9.13B). After removing that quantity of ice, is the Admiral standing at the water's edge (Figure 9.13C)? No—he stands on 9000 kg of ice (Figure 9.13D), because isostasy always readjusts the position of the iceberg so that nine-tenths of its mass is above the waterline. By digging downward (denuding the iceberg), Admiral Frost has caused the entire iceberg to move upward.

Recognition of isostatic uplift requires that we refine the vocabulary of uplift somewhat [38]. Uplift that truly increases the average elevation of the surface through time, without significant denudation at the same time, is called **surface uplift.** In contrast, **uplift of rocks** refers to any kind of upward vertical motion, including isostatic uplift, whether or not the elevation of the surface increases as a result. It has been suggested that most or all geomorphic and geodetic uplift measurements are actually measurements of uplift of rocks, incorporating both tectonic and isostatic driving forces. A geological test commonly used to infer amounts and rates of uplift is to collect rocks from the surface and determine the temperature and pressure and age at which those rocks formed. The temperature and pressure of formation yield an estimate of the depth of formation which, strictly speaking, is a measurement of **exhumation,** not uplift. Exhumation, like uplift of rocks, is distinct from surface uplift because it does not necessarily imply any increase in elevation of the surface [38]. The lesson for geologists, geodesists, and geomorphologists is to recognize the contributions of denudation and isostatic uplift, and direct their research to settings where the different processes can be isolated.

It is important to note that the **average elevation** of an area can only decrease as a result of isostatic uplift acting alone—the process *masks* the effects of downwearing, but it cannot *reverse* those effects. However, consider the case where erosion is not uniform across a mountain range, but rather is concentrated in the valley bottoms, increasing total relief. In this case, in the absence of tectonic uplift, average elevation decreases through time, but the **maximum elevation** of the region will increase (Figure 9.14). The high peaks of the range are isolated from most of the erosion below, but experience all of the isostatic uplift of the mountain block. Only when total relief in the range begins to decline will the maximum elevation also decline.

In mountain ranges, there are certain limitations on isostatic uplift [40]. Remember that the Earth's crust has strength, and isostatic compensation is always regional. For this reason, uplift of small geomorphic features, especially small warps such as described in Chapter 7, cannot be explained by isostasy. Furthermore, a number of authors have tried to invoke isostasy as a mechanism to explain episodic uplift and rejuvenation, but Gilchrist and Summerfield [41] have pointed out that isostasy operates continuously, without significant thresholds, and should not trigger episodic activity. However, Stüwe [42] outlines a credible exception to the rule whereby large-scale

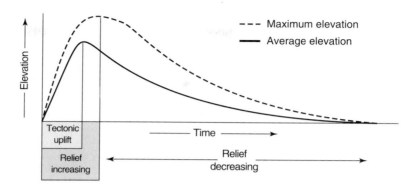

FIGURE 9.14
A brief pulse of tectonic uplift of a mountain range is combined with the long-term effects of erosion. The maximum elevation of the range continues to increase for a time after tectonic uplift ceases because, as long as *relief* is increasing, it triggers isostatic uplift.

retreat of escarpments on the African coast drives a repeating pattern of crustal flexure and renewed scarp retreat.

Equilibrium Landscapes

One of the most intriguing implications of denudation and isostasy is the possibility of an equilibrium landscape in a dynamic balance between uplift (tectonic and/or isostatic) and erosion. The best cases for real-world equilibrium mountain ranges are those with very high rates of uplift and precipitation. The Southern Alps form the sharp spine of South Island, New Zealand. Convergence across the Alpine fault that began around 2.5 Ma results in uplift rates as high as 22 mm/yr, and robust erosion is driven by rainfall of up to 11 m/yr at the range crest [43]. Erosion is inferred to equal uplift over the long term. The mountain range acts as an "erosional subduction zone," consuming crustal material that is fed into it by convergence on the Alpine fault.

In the Southern Alps, it appears that variations in rock type and structure, Quaternary climate change, vegetation, and human influence are all of secondary importance, swamped by the overwhelming influence of upheaval and erosion [43] (although this conclusion has been disputed [44]). It may turn out that other mountain ranges with lower rates of activity also may tend toward static, equilibrium forms, but the greater relative importance of the second-order effects listed above creates widespread variability and instability over shorter periods of time.

Another candidate for long-term landscape equilibrium is the topography of Japan. Uplift rates from numerous releveling surveys correlate strongly with denudation rates [36]. Using this quantitative relationship, Yoshikawa developed a model of mountain-range construction (Figure 9.15). One of the most interesting

FIGURE 9.15
Development of mean elevation of a mountain range by the simultaneous action of uplift and erosion. Note that in this model, ranges reach a stable maximum elevation, the value of which depends on the uplift rate.
(After Yoshikawa, 1985 [36])

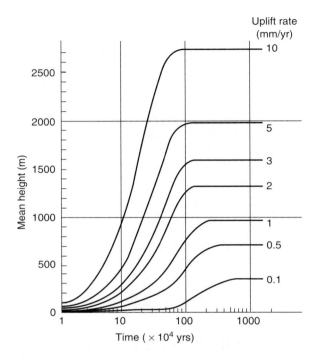

aspects of this model is that it predicts that, for a given climate and uplift rate, there is a maximum, dynamic elevation which a mountain range can reach. A range with a low uplift rate will reach its dynamic balance at a lower elevation than a range with more rapid uplift. According to this model, no amount of additional uplift will increase the elevation above the dynamic maximum. However, any increase in the *rate* of uplift will lead to a new, higher equilibrium elevation. Similarly, mountain ranges decay either when uplift ceases or when the rate decreases.

Are there characteristics common to all mountain ranges in a state of long-term dynamic equilibrium? The study of equilibrium landscapes is yet in its infancy, and the systems are so complex that no universal signature of equilibrium has been worked out. However, we are able to suggest some diagnostic traits. We expect these landscapes to be maturely dissected (see Figure 2.1); that is, with neither flat-topped plateaus nor deeply incised gorges. With all slopes graded to a mature drainage network, many peaks should be **accordant,** meaning that they are of about equal elevation or project to a uniform surface. Accordant summits have been recognized in the European Alps [45], in Japan [46], and elsewhere. Previously, patterns of accordant summits were interpreted to be the remnants of **peneplains** (low-relief erosional surfaces). But rather than being formed by an erosional base level, the peaks are now inferred to be the result of ongoing uplift.

Finally, we expect that an equilibrium landscape should demonstrate a pervasive relationship between topography and rock strength. In its simplest form, this means that the landscape will be higher and steeper over a durable bedrock than over weak rock. The Appalachian Mountains are the remnant of a mountain range that has been eroding for more than 100 m.y. They are a candidate for at least quasi-equilibrium ("quasi" because the only uplift is isostatic, and therefore elevations can only be reduced). The folded strata of the Valley and Ridge Province of the Appalachians (Figure 9.16) show a marked correlation between rock structure and topographic elevation. Interbedded strong and weak strata are folded into anticlines and synclines (Figure 9.17). The highest topography is on the limbs of the folds because, in this setting, erosion is controlled by the most resistant rock layers. The vertical thickness, and therefore the resistance, of a layer is greatest where the layer dips the most (Figure 9.17, inset), so those areas stand the highest.

THE PROBLEM WITH RATES

A central part of research in tectonic geomorphology, including mountain building, is determining rates of uplift. It should be possible to take rates of neotectonic

FIGURE 9.16
Satellite image of Valley and Ridge province of the Appalachians.
(Photo by Earth Satellite Corp./SPL/Photo Researchers, Inc.)

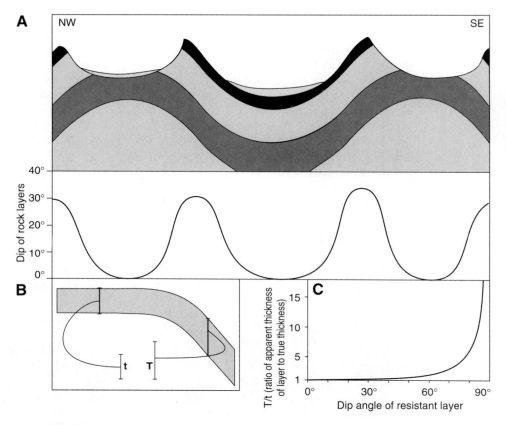

FIGURE 9.17

(A) Schematic illustration of relationship between structure and topography in the Valley and Ridge province of the Appalachians. (B) Note that, for any warped stratum, the vertical thickness is greatest where the layer dips the most steeply (inset). For a strong layer interbedded with weaker ones, the resistance to erosion is greatest where its thickness is greatest. (C) This fact explains why elevations are highest on the limbs of the folds.

activity and project them into the past and to compare them with rates from the longer geological record. However, at widely separated mountain ranges across the Earth, rates of Cenozoic activity (within the last 66 m.y.) appear to be several times faster than rates from the more distant geological past. Some researchers see this difference as a pulse of accelerated global tectonics [47], while others conclude that Cenozoic climate became much more intense and increased isostatic uplift worldwide [23].

This bias in uplift rates is not a new discovery. In 1949, James Gilluly, then the retiring president of the Geological Society of America, extolled the membership of the Society to reject a growing theory that most of geologic time has been

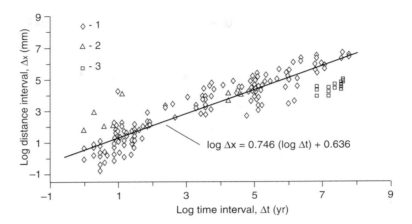

FIGURE 9.18
Log-log plot of uplift versus time interval for a large number of measured uplift rates. Different symbols indicate specific types of process: 1 - vertical crustal uplift, 2 - glacial crustal rebound and fault movement, 3 - salt or magma doming.
(From Gardner et al., 1987 [49])

anorogenic (without major tectonic activity), punctuated by brief pulses of tectonism, and that the present is one such atypical pulse [46]. He argued that the older geologic record is less complete, and the less complete the record, the more it will incorporate quiet times along with active periods. This bias in rates, sometimes called "the pull of the present" or "measurement interval bias," is not just limited to rates of uplift. Sedimentation [48], erosion [49], and evolution [50] all appear to have accelerated through the Cenozoic. Examining a large number of rates calculated from different intervals of time, the bias turns out to be quite systematic (Figure 9.18)—the longer the interval over which a rate is measured, the slower the apparent rate.

A statistical correction for this bias has been devised [49] in which a rate R_0 measured over a time interval Δt is corrected to a standardized yearly distance R_1

$$R_1 = R_0 \, (\Delta t^{\,0.255}) \qquad\qquad (9.19)$$

This correction allows comparison of uplift rates measured over intervals of time that differ by orders of magnitude. For example, a rate of mountain exhumation of 1.0 mm/yr calculated from a 10 Ma rock is equivalent to an uplift rate of 5.8 mm/yr calculated from a 10 ka stream terrace. The pull of the present is a demonstrable fact, but as long as the cause of the bias remains intangible, and as long as geologists can calculate an uncorrected rate on the back of a napkin, the myths of accelerated rates probably will continue.

TOWARD A COMPREHENSIVE CLASSIFICATION SCHEME

Knowing what we do about mountain ranges, it should be possible to characterize some of the differences and similarities between them. John Rodgers, introducing the subject of mountain-range anatomy, listed a number of specific characteristics: the nature of deformation, the particular assemblage of rock types, type of regional metamorphism, and other traits [51]. Rodgers also pointed out that all mountain ranges could be basically alike, but merely eroded to different levels. Imagine aliens trying to describe what a cow is like from different cuts of beef—they probably would describe a dozen different kinds of cow: a sirloin cow, a T-bone cow, a rib-eye cow, etc.

Various attempts have been made to create comprehensive classification models. In this chapter, we have emphasized the importance of the strength of uplift and of denudation. One model classifies different ranges in these terms [52]

1.   ranges where uplift > erosion
2.   ranges where uplift = erosion
3.   ranges where uplift < erosion

To this scheme, we would add the age of the mountain range, in order to distinguish between, for example, the 2.5 Ma Southern Alps, the 40–50 Ma Himalaya, and the 100 Ma Appalachians. Given information on these three parameters (uplift rate, erosion rate, and age), we should be able to make fundamental statements about a range: its elevation, relief, degree of exhumation, etc. The final step would be to integrate the quantitative relationships of uplift and erosion into unified models of landscape genesis. The first tentative steps have been taken in this direction, but it remains one of the great challenges of tectonic geomorphology today.

SUMMARY

Mountain ranges are the product of tectonic forces that cause uplift and erosional forces which sculpt the landscape. The effects of these processes can be seen at a variety of scales, from a single mountain peak to the entire planet. Once a mountain range has been constructed, it must also be supported. Possible support mechanisms include the buoyancy of the crust (isostasy) and the rigidity of the lithosphere (flexure). Erosion also plays an important role during and after construction of mountain ranges. In general, erosion rates increase with increasing elevation and relief, but rates have been quite variable through the climatic shifts of the last few millions of years. It is sometimes difficult to compare erosion rates with uplift rates because they are often measured over very different intervals of time, and "the pull of the present" has a significant effect on rates. However, in some locations, rates of uplift and erosion may be approximately equal, creating mountain ranges in dynamic equilibrium between creation and destruction.

317

REFERENCES CITED

1. Le Conte, J., 1909. *Elements of Geology*, Fifth Edition. D. Appleton and Company: New York.
2. Baker, V.R., 1986. Regional landform analysis. In N.M. Short and R.W. Blair, Jr. (eds.), *Geomorphology from Space: A Global Overview of Regional Landforms*. National Aeronautics and Space Administration: Washington, D.C.
3. Davis, W.M., 1899. The geographical cycle. *Geographical Journal*, 14: 481–504.
4. Penck, W., 1953. *Morphological Analysis of Landforms* (translated by E. Czech and K.C. Boswell). St. Martin's Press: New York.
5. Hack, J.T., 1960. Interpretation of erosional topography in humid temperate regions. *American Journal of Science*, 258-A: 80–97.
6. Balmino, G., K. Lambeck, and W.M. Kaula, 1973. A spherical harmonic analysis of Earth's topography. *Journal of Geophysical Research*, 78: 478–481.
7. Burchfiel, B.C., 1983. The continental crust. *Scientific American*, 249(3): 130–142.
8. Head, J.W., S.E. Yuter, and S.C. Soloman, 1981. Topography of Venus and Earth: a test for the presence of plate tectonics. *American Scientist*, 69: 614–623.
9. Saunders, R.S., R.E. Arvidson, J.W. Head III, G.G. Schaber, E.R. Stofan, and S.C. Soloman, 1991. An overview of Venus geology. *Science*, 252: 249–252.
10. Soloman, S.C., J.W. Head, W.M. Kaula, D. McKenzie, B. Parsons, R.J. Phillips, G. Schubert, and M. Talwani, 1991. Venus tectonics: analysis from Magellan. *Science*, 252: 297–312.
11. Head, J.W., 1990. Formation of mountain belts on Venus: evidence for large-scale convergence, underthrusting, and crustal imbrication in Freyja Montes, Ishtar Terra. *Geology*, 18: 99–102.
12. Vorder Bruegge, R.W., and J.W. Head III, 1991. Processes of formation and evolution of mountain belts on Venus. *Geology*, 19: 885–888.
13. James, D.E., 1973. The evolution of the Andes. In F. Press and R. Siever (eds.), *Planet Earth*. W.H. Freeman and Company: San Francisco.
14. Ramos, V.A., 1989. The birth of southern South America. *American Scientist*, 77: 444–450.
15. Mégard, F., 1987. Structure and evolution of the Peruvian Andes. In J.P. Scaer and J. Rodgers (eds.), *The Anatomy of Mountain Ranges*. Princeton University Press: Princeton, NJ.
16. Isacks, B.L., 1988. Uplift of the Central Andean Plateau and bending of the Bolivian orocline. *Journal of Geophysical Research*, 93: 3211–3231.
17. Molnar, P., 1986. The geologic history and structure of the Himalaya. *American Scientist*, 74: 144–154.
18. Williams S.H., and R.K. Stevens, 1974. Ancient continental margin of North America. In C.A. Burk and C.L. Drake (eds.), *Geology of Continental Margins*. Springer-Verlag: New York.
19. Murphy, J.B., and R.D. Nance, 1992. Mountain belts and the supercontinent cycle. *Scientific American*, 268(4): 84–91.
20. Molnar, P., 1989. The geological evolution of the Tibetan Plateau. *American Scientist*, 77: 350–360.
21. Harrison, T.M., P. Copeland, W.S.F. Kidd, and A. Yin, 1992. Raising Tibet. *Science*, 255: 1663–1670.
22. Sorkhabi, R.B., and E. Stump, 1993. Rise of the Himalaya: a geochronologic approach. *GSA Today*, 3: 85, 88–92.
23. Molnar, P., and P. England, 1990. Late Cenozoic uplift of mountain ranges and global climate change: chicken or egg? *Nature*, 346: 29–34.
24. Bull, W.B., 1977. The alluvial fan environment. *Progress in Physical Geography 1*, 222–270.
25. Bull, W.B., 1964. Geomorphology of segmented alluvial fans in western Fresno County, California. *U.S. Geological Survey Professional Paper* 352-E: 89–129.
26. Hooke, R.L., 1972. Geomorphic evidence of late-Winsconsin and Holocene tectonic deformation, Death Valley, California. *Geological Society of America Bulletin* 83, 2073–2098.

27. Wahrhaftig, C., 1970. Geomorphology (Chapter 7). In J. Verhoogen (ed.), *The Earth*. Holt, Rinehart and Winston Inc.: New York.

28. West, R.B., 1991. Tectonic geomorphology, landform modeling, and soil chronology of alluvial fans of the Tejon embayment, southernmost San Joaquin Valley, California. Master's thesis, University of California: Santa Barbara, California.

29. Pinter, N., and E.A. Keller, 1995. Geomorphic analysis of neotectonic deformation, northern Owens Valley, California. *Geologische Rundschau*, 84: 200–212.

30. Seaver, D., 1986. Quaternary evolution and deformation of the San Emigdio Mountains and their alluvial fans, Transverse Ranges, California. Master's thesis, University of California: Santa Barbara, California.

31. Gurnis, M., 1992. Long-term controls on eustatic and epeirogenic motions by mantle convection. *GSA Today*, 2: 141, 144–145, 156–157.

32. Molnar, P., 1986. The structure of mountain ranges. *Scientific American*, 255(1): 70–79.

33. Bechtel, T.D., D.W. Forsyth, V.L. Sharpton, and R.A.F. Grieve, 1990. Variations in effective elastic thickness of the North American lithosphere. *Nature*, 343: 636–638.

34. Holeman, J.N., 1968. Sediment yield of major rivers of the world. *Water Resources Research*, 4: 787–797.

35. Ahnert, F., 1970. Functional relationships between denudation, relief, and uplift in large mid-latitude drainage basins. *American Journal of Science*, 268: 243–263.

36. Yoshikawa, T., 1985. Landform development by tectonics and denudation. In A. Pitty (ed.), *Themes in Geomorphology*. Croom Helm: London.

37. Jackson, M., and 28 others, 1991. Trans-Himalayan geodesy. *EOS: Transactions, American Geophysical Union*, 72 (supplement): 112.

38. England, P., and P. Molnar, 1990. Surface uplift, uplift of rocks, and exhumation of rocks. *Geology*, 18: 1173–1177.

39. Bishop, P., and R. Brown, 1992. Denudational isostatic rebound of intraplate highlands: The Lachlan River valley, Australia. *Earth Surface Processes and Landforms*, 17: 345–360.

40. Pinter, N., and E.A. Keller, 1991. Comment on "Surface uplift, uplift of rocks, and exhumation of rocks" (England and Molnar, 1990). *Geology*, 19: 1053.

41. Gilchrist, A.R., and M.A. Summerfield, 1991. Denudation, isostasy, and landscape evolution. *Earth Surface Processes and Landforms*, 16: 555–562.

42. Stüwe, K., 1991. Flexural constraints on the denudation of asymmetric mountain belts. *Journal of Geophysical Research*, 96: 10,401–10,408.

43. Adams, J., 1985. Large-scale tectonic geomorphology of the Southern Alps, New Zealand. In M. Morisawa and J.T. Hack (eds.), *Tectonic Geomorphology*. Allen & Unwin: Boston.

44. Koons, P.O., 1989. The topographic evolution of collisional mountain belts: a numerical look at the Southern Alps, New Zealand. *American Journal of Science*, 239: 1041–1069.

45. Penck, A., 1919. Die Gipfelflur der Alpen. *Sitzung der Preussischen Akademie der Wissenschaften Berlin*, 17: 256–268.

46. Gilluly, J., 1949. Distribution of mountain building in geologic time. *Geological Society of America Bulletin*, 60: 561–590.

47. Kutzbach, J.E., W.L. Prell, and W.F. Ruddiman, 1993. Sensitivity of Eurasian climate to surface uplift of the Tibetan Plateau. *Journal of Geology*, 101: 177–190.

48. Sadler, P.M., 1981. Sediment accumulation and the completeness of stratigraphic sections. *Journal of Geology*, 89: 569–584.

49. Gardner, T.W., D.W. Jorgensen, C. Shuman, and C.R. Lemieux, 1987. Geomorphic and tectonic process rates: effects of measured time interval. *Geology*, 15: 259–261.

50. Gingerich, P.D., 1983. Rates of evolution: effects of time and temporal spacing. *Science*, 222: 159–161.

51. Rodgers, J., 1987. Differences between mountain ranges. In J.P. Scaer and J. Rodgers (eds.), *The Anatomy of Mountain Ranges*. Princeton University Press: Princeton, NJ.

52. Isacks, B.L., 1992. 'Long-term' land surface processes: erosion, tectonics and climatic history in mountain belts. In P.M. Mather (ed.), *TERRA-1: Understanding the Terrestrial Environment, The Role of Earth Observations from Space*. Taylor & Francis: London.

Appendix A:
Selected Dating Methods

Numerical Dates — Annual / Radiometric

Method	Applicability	Age Range and Optimum Resolution					Basis of Method and Remarks
		10^2	10^3	10^4	10^5	10^6	
Annual							
Historical records	X to XXX	=====+===					Requires preservation of pertinent records; applicability depends on quality and detail of records. Limited to several hundred years in western hemisphere.
Dendrochronology	XX	=======					Requires either direct counting of annual rings back from present or construction of a chronology based on variations in annual ring growth. Restricted to areas where trees of the required age and (or) environmental sensitivity are preserved.
Varve chronology	X	=======+++					Requires either direct counting of varves back from present or construction of a chronology based on overlapping successions of continuous varved lake sediments. Subject to errors in matching separate sequences and to misidentification of annual layers.
Radiometric							
Carbon-14	X to XXX	o•-+++=====?					Depends on availability of carbon. Based on decay of ^{14}C, produced by cosmic radiation, to ^{14}N. Subject to errors due to contamination, particularly in older deposits and in carbonate material (such as mollusk shells, marl, soil carbonate).
Uranium series	XX			••- - - -++++-			Used to date coral, mollusks, bone, cave carbonate, and carbonate coats on stones. Potentially useful in dating travertine and soil carbonate. A variety of isotopes of the U-decay series are used during $^{230}Th/^{234}U$ (most common and method described to left), $^{234}U/^{238}U$ (with a range back to 600,000 yr), $^{231}Pa/^{235}U$ (10,000 –12,000 yr), U-He (0–2 m.y.), and $^{226}Ra/^{230}Th$ (<10,000 yr). Errors due to the lack of a closed chemical system are a common problem, especially in mollusks and bone.
Potassium-argon	X			••••- - -+++++			Directly applicable only to igneous rocks and glauconite. Requires K-bearing phases such as feldspar, mica, and glass. Based on decay of ^{40}K to ^{40}Ar. Subject to errors due to excess argon, loss of argon, and contamination.

Numerical Dates (cont.) — **Radiometric (cont.)**

Method	Applicability	Age Range and Optimum Resolution (10^2, 10^3, 10^4, 10^5, 10^6)					Basis of Method and Remarks
Fission track	X			•••–—-+++			Directly applicable only to igneous rocks (including volcanic ash); requires uranium-bearing material (zircon, sphene, apatite, glass). Based on the continuous accumulation of tracks (strained zones) caused by recoiling U fission products. Subject to errors due to track misidentification and to track annealing.
Uranium trend	XXXX		∞∞••——––••				Based on open-system flux of uranium through sediment and soil; ^{238}U, ^{234}U, ^{230}Th, and ^{232}Th must be measured on about five different samples from a given-aged deposit and an isochron constructed to determine age.
Thermoluminescence (TL) and electron-spin resonance (ESR)	XXXX		∞∞••••••————————••				Based on displacement of electrons from parent atoms by alpha, beta, and gamma radiation. Applicable to feldspar and quartz in sediments and carbonate in soils. TL based on amount of light released as sample is heated compared with that released after known radiation dose. TL precision better than indicated for ceramics in 400–10,000-year range.
Cosmogenic isotopes other than carbon–14	X	–?–?–?–?–?–?–?–?–?					Dating methods analogous to ^{14}C-dating are based on the cosmogenic isotopes (half-life in years in parentheses), ^{32}Si (300), ^{41}Ca (1.3×10^5), ^{36}Cl (3.08×10^5), ^{26}Al (7.3×10^5), ^{10}Be (1.5×10^6), ^{129}I (1.6×10^7), and ^{53}Mn (3.7×10^6).

Relative Dates

Method	Applicability	Age Range					Basis of Method and Remarks
Amino acid racemization	XX	•••••••••••••——–•••					Requires shell or skeletal material. Based on release of amino acids from protein and subsequent inversion of their stereoisomers. Shells tend to be more reliable than bone, wood, or organic-rich sediment. Is strongly dependent on other variables, especially temperature and leaching history. Commonly used as a relative dating or correlation technique, but yields numerical ages when calibrated by other techniques.
Obsidian hydration	X	————————•••					Based on thickness of the hydrated layer along obsidian crack or surface formed during given event. Age proportional to the thickness squared. Calibration depends on experimental determination of hydration rate or numerical dating. Subject to errors due to temperature history and variation in chemical composition.
Tephra hydration	X	∞∞∞∞∞∞∞••••••••					Requires volcanic ash. Based on the progressive filling of bubble cavities in glass shards with water. Subject to the same limits as obsidian hydration, plus others, including the geometry of ash shards and bubble cavities.
Lichenometry	X to XXX	————••••					Requires exposed, stable rock substrates suitable for lichen growth. Most common in alpine and arctic regions, where lichen thallus diameter is proportional to age. Subject to error due to climatic differences, lichen kill, and misidentification. The limit of the useful range varies considerably with climate and rock type.

	Method	Applicability	Age Range and Optimum Resolution					Basis of Method and Remarks
			10^2	10^3	10^4	10^5	10^6	
Relative Dates (cont.)	Soil development	XXXX	ooooo•••••• — — — — — ••†					Encompasses a number of soil properties that develop with time, all of which are dependent on other variables in addition to time (parent material, climate, vegetation, topography). Is most effective when these other variables are held constant or can be evaluated. Precision varies with the soil property measured; for example, accumulation of soil carbonate locally yields age estimates within ±20 percent.
	Rock and mineral weathering	XX	ooooo ••••••—————————•••					Includes a number of rock- and mineral-weathering features that develop with time, such as thickness of weathering rinds, solution of limestone, etching of pyroxene, grussification of granite, and buildup of desert varnish. Has the same basic limitations as soil development. Precision varies with the weathering feature measured.
	Progressive landform modification	XXX	ooooooooo•••••••••ooo					In addition to time, depends on factors such as climate and lithology. Depends on reconstruction of original landform and understanding of process resulting in change of landform, including creep and erosion.
	Rate of deposition	XX	— —?— —?— —?— —?—?— —					Requires relatively constant rate of sedimentation over time intervals considered. Numerical ages based on sediment thickness between horizons dated by other methods. Quite variable in alluvial deposition.
	Geomorphic position and incision rate	XXX	ooooooooo ••••••• —?— —?—					Geomorphic incision rates depend on stream size, sediment load, bedrock resistance to erosion, and uplift rates or other base-level changes. If one terrace level is dated, other terrace levels may be dated assuming constant rate of incision.
	Rate of deformation	XXX	••?••?••?••? ••?••? ••?••?••?					Dating assumes deformation rate constant over interval of concern and requires numerical dating for calibration. At spreading centers and plate boundaries, nearly constant rates may be valid for intervals of millions of years.
Correlation	Stratigraphy	XXXX						Based on physical properties and sequence of units, which includes superposition and inset relations. Depends on the establishment of time equivalence of units; deposition of Quaternary units normally occurs in response to cyclic climatic changes.
	Tephrochronology	X	RESOLUTION DEPENDS ON RECOGNITION OF FEATURE AND ACCURACY OF DATING THAT FEATURE					Requires volcanic ash (tephra) and unique chemical or petrographic identification and (or) dating of the ash. Very useful in correlation because an ash eruption represents a virtually instantaneous geologic event.
	Paleomagnetism	XX						Depends on correlation of remnant magnetic vector, which includes polarity, or a sequence of vectors with a known chronology of magnetic variation. Subject to errors due to chemical magnetic overprinting and physical disturbance.

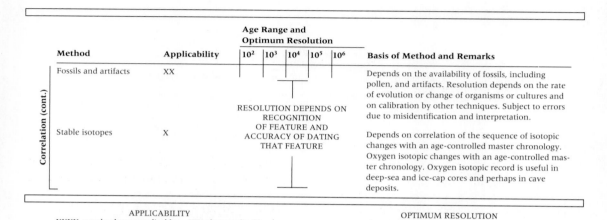

	Method	Applicability	Age Range and Optimum Resolution					Basis of Method and Remarks
			10^2	10^3	10^4	10^5	10^6	
Correlation (cont.)	Fossils and artifacts	XX						Depends on the availability of fossils, including pollen, and artifacts. Resolution depends on the rate of evolution or change of organisms or cultures and on calibration by other techniques. Subject to errors due to misidentification and interpretation.
	Stable isotopes	X			RESOLUTION DEPENDS ON RECOGNITION OF FEATURE AND ACCURACY OF DATING THAT FEATURE			Depends on correlation of the sequence of isotopic changes with an age-controlled master chronology. Oxygen isotopic changes with an age-controlled master chronology. Oxygen isotopic record is useful in deep-sea and ice-cap cores and perhaps in cave deposits.

APPLICABILITY

XXXX, nearly always applicable   XX, often applicable

XXX, very often applicable   X, seldom applicable

OPTIMUM RESOLUTION

======, <2 percent   ••••••, 25–75 percent

++++++, 2–8 percent   oooooo, 75–200 percent

––––––, 8–25 percent

(Modified slightly from Pierce, K.L., 1986. Dating methods. In *Active Tectonics*)

Glossary

accelograph: instrument that measures ground acceleration during seismic shaking.

accordant summits: a landscape in which all the peaks have approximately the same elevations.

active fault: a fault that has moved within a given period of time, typically the last 10,000 years.

aftershock: an earthquake that follows and is less powerful than the main shock.

aggradation: in a *fluvial system,* the accumulation of sediments in response to a rise in *base level* or other causes.

alignment (leveling) array: line of *control points* used to measure displacement across a fault.

alluvial fan: a conical, depositional landform found along many mountain fronts of arid and semiarid regions.

alluvium: loose sedimentary material deposited by rivers or streams.

antecedent stream: a stream that existed before geologic processes altered the landscape; such streams commonly cut through geologic structures or across high-standing topography of the modern landscape.

antithetic shear: a secondary fault with the opposite sense of displacement as the main fault.

Archimedes' principle: principle that the weight of a floating solid equals the weight of fluid that it displaces.

aseismic: describes an event or process that occurs without accompanying earthquake activity.

asthenosphere: plastic layer of the upper mantle that lies beneath the *lithosphere.*

asymmetry factor (AF): a geomorphic index used to detect active tilting.

balanced cross section: a geologic cross section in which strata are parallel, and individual layers maintain uniform or uniformly-varying thickness.

base level: the lowest elevation that a specific fluvial system drains to; the concept includes both *local base level* and *ultimate base level* (usually sea level).

beheaded stream: a stream that ends upstream at a fault and has been faulted away from its headwaters.

body waves: seismic waves that travel through the interior of the Earth.

brittle behavior: when a material responds to an applied stress by fracturing.

buried reverse fault: a compressional fault that does not or did not break the ground surface when it was last active.

capable fault: according to the U.S. Nuclear Regulatory Commission definition, a fault that has moved at least once in the past 50 k.y. or more than once in the past 500 k.y.

catastrophe: a disaster from which recovery is a long and involved process.

characteristic earthquake: an earthquake that strikes a given fault zone with approximately the same magnitude and with similar characteristics at approximately equal intervals.

colluvial wedge: a deposit of colluvium at the base of a slope, thickest near the slope and progressively thinner farther away.

colluvium: loose sedimentary material deposited by gravity-driven processes, usually at the base of a slope.

complex response: a model that states that features within a system (such as individual landforms in a landscape system) may be caused by changes in *intrinsic variables,* and not the direct result of external stimuli.

conditional probability: probability of seismic risk based on available knowledge.

continental drift: movement of continents in response to sea-floor spreading. The most recent episode of continental drift began about 200 million years ago with the breakup of the supercontinent Pangaea.

control points: surveyed points in a geodetic net or array.

convergent plate boundary: boundary between two lithospheric plates in which one plate descends below the other (subduction).

coseismic: describes an event or process that coincides with an earthquake.

creep: see *tectonic creep*.

critical facilities: facilities that, if damaged or destroyed, might cause catastrophic loss of life, property damage, or disruption of society.

Cycle of Erosion: model of landscape development developed by W.M. Davis. In this model, landscapes go through three characteristic stages: youth, maturity, and old age.

collement: a low-angle structure (typically a *thrust fault*) separating more

deformed rocks above from less deformed rocks below.

deflected drainage: a stream that follows a strike-slip fault along some or all of its length.

degradation: in a *fluvial system,* the removal of sediments or erosion of the channel in response to a fall in *base level* or other causes.

dendritic drainage pattern: "finger-like" pattern of streams associated with homogeneous bedrock and gentle slopes.

dendrochronology: the study of tree rings.

denudation: regional erosion of the surface.

depth of compensation: in *isostasy*, it is the depth in a fluid at which pressure is everywhere equal.

detachment fault: a low-angle normal fault across which there is significant displacement.

diffusion equation: a mathematical expression that is used to quantitatively model *fault-scarp degradation.*

dilatancy: the development of cracks and pores in a material that is subjected to stress.

dilatancy-diffusion model: the theory that dilatancy in rocks near a fault zone leads to an influx of water that triggers an earthquake.

dip: the maximum slope angle on a sloping surface.

divergent plate boundary: boundary between lithospheric plates characterized by production of new lithosphere; found along oceanic ridges.

drainage basin: the area in which all rain that falls exits through the same stream.

drainage density: the number and length of channels per unit area.

drape fold: a fold that forms over, and as a result of, a buried normal fault.

dry-tilt net: three *control points* used to measure tilt of the ground surface.

dynamic equilibrium: a condition of stability that is created by self-adjustment of all processes operating within the system (see *intrinsic variables* and *threshold*).

earthquake: a sudden motion or trembling in the Earth caused by the abrupt release of strain on a *fault*.

Earthquake Cycle: model in which earthquakes are the result of the accumulation of elastic strain.

earthquake precursors: events or phenomena that precede an earthquake.

elastic behavior: deformation that is recovered fully and instantaneously when the driving force is removed.

elastic thickness: in *isostasy*, it is the depth to which the crust acts as a rigid or brittle solid.

emergence: motion of the land up relative to sea level, such that the coastline advances oceanward through time.

ephemeris: a mathematical model of the precise orbital path of a satellite.

epicenter: the point on the surface of the Earth directly above the *focus* of an earthquake.

erosion: general term describing the processes of *weathering* and transport of sediment.

exhumation: the unburial of rocks by erosion.

extrinsic variables: processes that originate or operate outside of a system.

fault: a fracture or fracture system that has experienced movement along opposite sides of the fracture.

fault-bend fold: a fold formed by a change in dip on an underlying fault.

fault gouge: a clay zone formed by pulverized rock during an earthquake, which may create a groundwater barrier.

fault-propagation fold: a fold that forms around the tip of a fault that does not rupture the ground surface (see *buried reverse fault*).

fault scarp: a steep slope formed by a fault rupturing the surface.

fault scarp degradation: erosion of a fault scarp. See *morphological dating.*

fault-valve mechanism: hypothesis that earthquakes may be closely linked with fluid pressures in the crust, and that discharges of crustal fluids may accompany earthquakes.

fault zone: a related group of faults in a subparallel belt or zone.

fill terrace: a type of *stream terrace* that consists of a thick accumulation of *alluvium* (in contrast to a *strath terrace*).

flexural-slip fault: a fault on a stratigraphic bedding plane or other plane of weakness caused solely by flexure associated with active folding.

flexural support: a process by which the weight of mountains or other loads atop the Earth's crust are carried by the strength of the crust.

flexure: bending of a material or a surface (see *warping*).

floodplain: flat land or valley floor that borders a stream or river, formed by migration of meanders and/or periodic flooding.

fluvial geomorphology: the study of river processes and landforms caused by river processes.

fluvial system: a river or stream.

focus: the location within the Earth at which an earthquake originates.

fold-and-thrust belt: a zone characterized by subparallel faults and folds that reflect active compression.

footwall: the side of a fault that lies beneath the inclined fault plane.

foreshock: an earthquake that precedes and is less powerful than the main shock.

geodesy: study and measurement of the shape of the Earth's surface.

geoid: a surface of equal-gravity potential around the Earth. In most applications, mean sea level is used.

geomorphic record: the sum of landforms and *Quaternary* deposits at a site or in an area.

geomorphology: the study of landforms and surface processes.

Global Positioning System (GPS): geodetic positioning technique that utilizes the array of satellites maintained by the U.S. Department of Defense.

graben: a structural block that is down-dropped by faults on both sides of it.

graded river: a river in which driving forces (e.g., gravity, slope, discharge) and resisting forces (e.g., volume of sediment transported, channel roughness, sinuosity) are equal along its entire length.

ground acceleration: a quantitative measurement of the intensity of seismic shaking (usually given as a percent of the acceleration of gravity).

ground-penetrating radar: a method of sending radar waves into the subsurface and measuring their reflections with the goal of determining geometry beneath the surface.

half graben: a structural block that is downdropped by a fault on only one side of it. Subsidence of a half graben must be accompanied by *tilting* or *warping*.

hanging wall: the side of a fault that lies above the inclined fault plane.

Holocene Epoch: the last 10,000 years.

horst: a structural block that is lifted up relative to the blocks on either side of it.

hypocenter: the point in the earth where an earthquake originates; also known as the *focus*.

hypsometric curve: a graphical representation of the elevation distribution of a given landscape.

hypsometric integral: area under the *hypsometric curve*.

hypsometry: measurement and analysis of the distribution of land area at different elevations.

incision: local or regional erosion by streams, typically causing an increase in *relief*.

induced seismicity: earthquakes caused by human activity such as building dams, injecting fluids into the subsurface, or underground testing of nuclear weapons.

intensity (of an earthquake): a relative measurement of the strength of shaking at any given location. Intensity generally decreases with increasing distance from the epicenter.

interseismic: describes an event or process that occurs between major earthquakes.

intraplate earthquakes: earthquakes that occur in the interior of a lithospheric plate, away from any plate boundary.

intrinsic variables: processes that originate or operate within a system.

isostasy: the principle by which thicker, more buoyant crust stands topographically higher than thinner, denser crust.

landform: a discrete element of the landscape, such as a hill, a terrace, or an alluvial fan.

landslide: any downslope motion under the force of gravity—sometimes a secondary effect of earthquake shaking.

liquefaction: transformation of water-saturated sediments from a solid to a liquid state in response to shaking.

listric fault: a *normal fault* that is curved.

lithosphere: the upper portion of the Earth, consisting of the crust and upper portion of the mantle, that is characterized by brittle behavior.

locked fault: a fault which does not exhibit *tectonic creep*; a fault on which stress accumulates.

longshore transport: movement of sediment parallel to the shoreline as a result of waves that strike the shore at an oblique angle.

Love wave: a type of surface wave that causes a sideways shaking (in contrast to *Rayleigh wave*).

magnitude (of an earthquake): an absolute measurement of the energy of a given earthquake.

marine terrace: (sometimes called an *uplifted marine terrace*) a set of coastal landforms, typically either a *wave-cut platform* or a coral reef complex, that has been uplifted above the modern shoreline.

material amplification: a local increase in the intensity of seismic shaking caused by near-surface material (usually loose sediments or artificial fill).

maximum credible earthquake: the largest earthquake likely to be generated by faults in a given area.

meander: one of a series of curves in a sinuous stream or river.

measurement interval bias: observation that the rates of many processes seem to be slower in the distant geologic past than they are in the present or in the recent past.

mid-ocean ridge: divergent spreading center at the center of an ocean, called a ridge because the newly formed ocean crust is relatively buoyant and causes the topography to stand high.

Modified Mercalli Scale: a system for estimating earthquake *intensity.*

moment magnitude: a system for measuring earthquake *magnitude* based on the total energy released by the earthquake (also see *seismic moment*).

monsoon: pattern of seasonal storms that strike India, Southeast Asia, western Africa, and northern Australia.

morphological dating: estimating the age of a landform based on its shape, usually estimating the amount of erosion that has occurred since the landform was formed.

morphometry: quantitative measurement and analysis of topography.

mountain front: steep escarpment that marks the boundary between mountainous topography and relatively flat topography.

mountain-front sinuosity (S_{mf}): a geomorphic index used to detect tectonic activity along mountain fronts.

normal fault: a fault across which there is extension.

numerical age control: estimates of the age of a material or feature in an absolute number of years (as opposed to *relative age control*).

offset stream: a stream the channel of which is displaced across a strike-slip fault.

orogeny: regional increase in topography caused by tectonic processes (mountain-building).

orthometric height: elevation above or below the geoid.

P-wave: compressional ("push-pull") *seismic waves.*

paleoseismology: the study of earthquakes that occurred in the geologic past.

peneplain: a low-relief plain that is the theoretical end-product of erosion without tectonic activity (see *Cycle of Erosion*).

piercing points: two points on opposite sides of a fault that were originally connected, but were offset by one or more ruptures along the fault.

plastic behavior: a permanent change in the shape of a material after a force is applied to it.

plate tectonics: a model of global tectonics that suggests that the outer layer of the earth known as the lithosphere is composed of several large plates that move relative to one another; continents and ocean basins are passive riders on these plates.

Pleistocene Epoch: the period of geologic time from about 2 million to 10,000 years ago. Much of the Pleistocene was characterized by the growth and decline of glaciers in many areas.

post-glacial rebound: uplift caused by *isostasy* that follows the melting of a large continental ice sheet.

postseismic: describes an event or process that occurs shortly after an earthquake.

preseismic: describes an event or process that occurs shortly before an earthquake.

pressure ridge: a hill along a strike-slip fault zone formed by upwarping at *restraining bends* or between two different strands of the fault.

process-response model: a model of the development of landforms and the landscape based on an understanding of the processes at work and how they shape the surface.

pull of the present: see *measurement interval bias.*

Quaternary: the latest period of geologic time up to and including the present. The Quaternary includes the *Pleistocene* and the *Holocene,* and ranges from approximately 1.65 million years ago to the present.

radiocarbon dating: a method that estimates the absolute age of a sample based on the ratio of radiogenic carbon (^{14}C) to stable carbon (^{12}C and ^{13}C).

Rayleigh wave: a type of surface wave that causes an elliptical motion like the rolling of an ocean wave (in contrast to *Love wave*).

recurrence interval: the average period of time between major earthquakes on a given fault (see *characteristic earthquake*).

rejuvenation: renewed uplift and erosion of a mature landscape (see *Cycle of Erosion*).

relative age control: estimates of the age of a material or feature compared with other features (as opposed to *numerical age control*).

relative spacing: a method for determining the ages of a sequence of *marine terraces* based on their present-day elevations and knowledge of Quaternary sea-level history.

relative tectonic activity class designation: classification of a given area into one of several categories of ongoing tectonic activity.

releasing bend: a bend in a strike-slip fault that causes extension across the area of the bend.

relief: generally, the "ruggedness" of the topography; specifically, the highest elevation minus the lowest elevation in a given area.

response spectra: in earthquake engineering, it is the relationship between seismic-wave period and ground shaking.

restraining bend: a bend in a strike-slip fault that causes compression across the area of the bend.

retrodeformation: interpretation of a geologic cross section with the goal of understanding its geometry before deformation occurred.

reverse fault: a fault across which there has been convergence.

Richter magnitude: (also called *local magnitude*) a system for measuring earthquake *magnitude* based on the maximum amplitude and period of seismic waves recorded by a seismograph at a set distance from the earthquake *epicenter.*

rupture: breakage of a material under stress.

S-wave: *seismic waves* in which particle displacement is perpendicular to the direction of propagation of the wave.

sag pond: a pond along a strike-slip fault zone formed by downwarping between two different strands of the fault.

sand boil: sand extruded from the surface during seismic shaking, caused by high fluid pressure and *liquefaction*. Also called a "sand blow" or "sand volcano."

Satellite Laser Ranging (SLR): geodetic positioning technique based on the travel time of a laser pulse from a measuring station to a ranging satellite back to the measuring station.

satellite radar interferometry: geodetic positioning technique that uses pairs of radar images with the goal of measuring small changes in position over broad areas.

scarp: (short for escarpment) a slope steeper than the surrounding topography; related to a change in material, process, or geomorphic history.

seacliff: on a *marine terrace* or a modern coastline, it is the steep slope cut by wave action at its base.

segmentation: subdivision of a fault zone into smaller units with discrete rupture histories.

seismic: refers to vibrations in the earth produced by earthquakes.

Seismic Deformation Cycle: a repeating pattern of *preseismic, coseismic, postseismic,* and *interseismic* deformation at or near a fault.

seismic gap: a portion of a fault zone, between two areas that have ruptured in historical or recent time, that has *not* ruptured.

seismic moment: a measurement of the total amount of energy released during an earthquake.

seismic reflection profiling: a method of inducing vibrations and measuring their reflections with the goal of determining geometry beneath the surface.

seismic risk: an estimate of the likelihood and the potential damage of an earthquake in a given area.

seismic waves: energy released from a fault rupture, subdivided into *body waves* (*P-waves* and *S-waves*) and *surface waves* (*Rayleigh waves* and *Love waves*).

seismic zoning: legal definition of land as appropriate or inappropriate for different uses based on proximity to active faults, presence of material that may amplify shaking, etc.

seismograph: instrument for measuring *seismic waves;* the record of seismic waves itself is called a *seismogram.*

seismometer: instrument used to measure seismic waves.

shoreline angle: on an erosional coastline, the line at which the wave-cut platform meets the seacliff.

shutter ridge: a ridge offset by a strike-slip fault such that the ridge below the fault is juxtaposed against the gully above the fault.

slip rate: long-term rate of motion on a fault or fault zone.

soil chronosequence: a series of soil profiles systematically arranged from youngest to oldest.

strain: deformation resulting from *stress.*

strain partitioning: observation that in areas with oblique strain, horizontal and vertical deformation often occur on distinct and separate structures.

strath terrace: a type of *stream terrace* that consists of a cut bedrock surface with little or no *alluvium* overlying it (in contrast to a *fill terrace*).

stream length-gradient index (SL): a geomorphic index used to identify possible areas of tectonic activity.

stream power: the rate at which a stream can do work.

stress: force per unit area; may be compressive, tensile, or shear.

strike: the orientation of a horizontal line on a sloping surface.

strike-slip fault: a fault across which displacement is predominantly horizontal.

subduction: process in which one lithospheric plate descends beneath another.

subduction zone: a convergent plate boundary at which dense crust of one of the plates sinks down into the mantle.

subenvelope map: a topographic map based on stream elevations, with the higher elevations between streams stripped away.

submergence: motion of the land down relative to sea level, such that the coastline advances landward through time.

subsidence: downwarping of an area of the Earth's surface.

surface uplift: a specific type of *uplift* in which the elevation of the surface increases through time.

surface waves: seismic waves that travel along the surface of the Earth or discrete boundaries within the Earth.

synthetic shear: a secondary fault with the same sense of displacement as the main fault.

tectonic creep: movement along a fault zone that does *not* occur at the same time as an earthquake.

tectonic cycle: the global process of the creation, evolution, and destruction of the crust of the Earth.

tectonic geomorphology: (1) the study of landforms shaped by tectonic process; (2) application of geomorphic principles to reveal the presence, pattern, or rates of tectonic processes.

tectonics: processes of deformation (motion) of the Earth's crust, and the structures and landforms that result from those processes.

terrace: an inactive bench, typically near an active stream or coastline. The term is applied to both the flat surface of the terrace (the *tread*) and the slope below (the *riser*).

threshold: a critical transition point, such as the maximum amount of change that a system can absorb before its *dynamic equilibrium* becomes unbalanced.

thrust fault: a type of *reverse fault* which is less steep than 45°.

tide gauge: an instrument used to measure sea level over time.

tilting: process by which a horizontal surface acquires a slope (usually without *warping*).

time history: properties of ground shaking over the duration of an earthquake at a particular site.

transform fault: type of fault associated with oceanic ridges; may form a plate boundary, such as the San Andreas Fault in California.

transport-limited slope: a hillslope on which erosion is limited only by the rate of sediment transport (also see *weathering-limited slope*).

transverse topographic symmetry factor (T): a geomorphic index used to detect active tilting.

trilateration net: network of *control points* used to measure vertical or horizontal deformation.

triple junction: the point at which three plate boundaries meet.

tsunami: a potentially damaging ocean wave triggered by a submarine earthquake or other large-scale shift on the ocean floor.

uniform-slip model: a model in which a fault or faults are characterized by constant displacement per rupture event, a constant long-term slip rate, and a characteristic earthquake magnitude.

unpaired stream terrace: a terrace segment on only one side of the modern stream; less likely to reflect a regional tectonic or climatic event than *paired terraces*.

uplift of rocks: any upward vertical motion, whether or not the elevation of the surface increases as a result.

uplift path: on a graph of sea-level history (elevation versus age), it is the line that traces how an individual *marine terrace* was formed and uplifted to its present elevation.

variable-slip model: a model in which a fault or faults are characterized by irregular amounts of slip per rupture event, long-term slip rates, and/or earthquake magnitudes.

vertical extent of mortality (VEM): vertical width of a zone in which non-mobile marine organisms are killed by coseismic uplift.

Very Long Baseline Interferometry (VLBI): geodetic positioning technique that uses radio telescopes to measure signals from quasars.

volcanic tumescence: uplift caused by rising magma beneath the surface.

warping: process by which a planar surface becomes folded.

wave-cut platform: on a *marine terrace* or a modern coastline, the subhorizontal surface cut by waves as well as by secondary biological and chemical processes.

wavelength: in *isostasy* and *flexural support,* the lateral dimensions of a load on the Earth's crust.

weathering: the sum of all physical and chemical processes that break down rock at and near the Earth's surface.

weathering-limited slope: a hillslope on which erosion is limited only by the rate of bedrock (or substrate) weathering (also see *transport-limited slope*).

winter berm: (also called the *storm berm*) on a sandy shoreline, it is the high ridge of sand that marks the highest limit of wave action during winter storms (also see *beach ridges*).

Index

Bold page numbers indicate figures.

abrupt burial, 246, 266
acceleration, 40, **42,** 315
accelograph, 40
accommodation zone, 71
accordant summits, 312
aftershock, 37
afterslip, 111
aggradation, 151, 159, 167, 258
Alaska ("Good Friday") earthquake (1964), 30, **32, 34,** 190
alignment arrays, 98
alluvial fans, 52, 55, 64, **65,** 141, 144, 239, 299–**302**
Alpine Fault, New Zealand, 199, 311
amplifiction, 21–22, 288
amplitude, 13
Andes Mountains, 294–295
Anti-Spoofing, 108
Apennine Mts., Italy, 79
Appalachian Mts., 129, 297, **314,** 316
Appalachian Valley, 128–129
aseismic deformation. *See* creep
asthenosphere, 5, 305

asymmetry factor, 126–128, **127**
attenuation, 21
average peak horizontal acceleration, 18
avulsion, 151

base level, 123, 158–159, 171, 293
Basin and Range province, 71, **74, 75,** 115, 256, 270, 303
beaches, **186, 187**–197
benches, 62
bend
  releasing, 65, **67**
  restraining, 65, **67**
bioerosion platform, 183–**184**
Blue Ridge, 129
Borah Peak, Idaho earthquake (1983), 2, 280
boreholes, 248
bracketed duration. *See* duration of shaking
braided channel, 57
buried reverse faults, 9, **11,** 173, 212, 240
buried thrust fault. *See* buried reverse faults

Cape Mendocino earthquake (1992), 179, **181**–190

Cascadia subduction zone, 179, 192, 212–**213, 252**
catastrophes, 2, 3, 26, 31, 192, 203, 276, 281
characteristic earthquake, 37, 269, 271
Charleston, South Carolina earthquake (1886), 153
China, 228
chronology, 82, **83**–87, 239, 257, 264
^{14}C method, **83**–85, 160, 257
Coalinga, California earthquake (1983), 212
coastlines, 181–205
  carbonate, 182, **188**
  classification of, 182–**183**
  clastic, 182, 185
  erosional, 182–**185**
  marine terraces (*see* terraces, marine)
colluvial wedges, 184, 246, 262–**263**
Colorado River, 307
complex response, 53, 55–57, 170–171
conditional probability, 27, **28, 29,** 268, 274–275
continental collision, 5, **296, 298**
coral reefs, 179, 185–186, **189, 194,** 196, 202